MOON

From 4.5 billion years ago to the present

COVER IMAGE: G.D. Cassini's 1679 map of the Moon.

Dedication

To Paul Spudis, ever helpful.

First published in February 2016
Reprinted May 2019

A catalogue record for this book is available from the British Library.

ISBN 978 0 85733 826 6

Library of Congress control no. 2015948113

Published by Haynes Publishing,
Sparkford, Yeovil,
Somerset BA22 7JJ, UK.
Tel: 01963 440635
Int. tel: +44 1963 440635
Website: www.haynes.com

Haynes North America Inc.,
859 Lawrence Drive, Newbury Park,
California 91320, USA.

Printed in Malaysia.

Acknowledgements

I would like to thank Steve Rendle of Haynes, James Joel Knapper, Ken MacTaggart, Gordon Seales, Philip Stooke and Don Wilhelms. W. D. Woods provided sterling assistance in checking the draft and preparing many of the illustrations. Unless otherwise stated, an illustration is in the public domain.

MOON

From 4.5 billion years ago to the present

Owners' Workshop Manual

An insight into the study and exploration of
our nearest celestial neighbour

David M. Harland

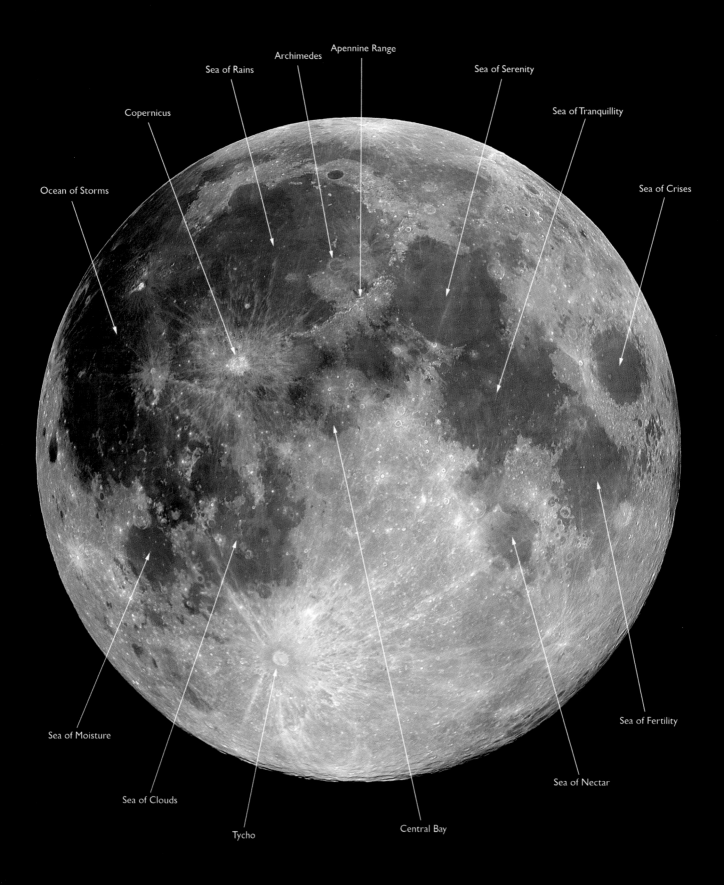

Archimedes

Apennine Range

Sea of Rains

Copernicus

Sea of Serenity

Sea of Tranquillity

Ocean of Storms

Sea of Crises

Sea of Moisture

Sea of Clouds

Tycho

Central Bay

Sea of Nectar

Sea of Fertility

Contents

OPPOSITE Some of the lunar features on the near side that are mentioned in the text. *(Woods)*

Introduction

This manual reviews how our understanding of the Moon has developed.

The story starts with a brief account of how science and in particular astronomy were studied in classical times. The dominant figure was the philosopher Aristotle, who stated that Earth was located at the centre of the universe and that objects in the celestial realm pursued uniform and circular motions.

Very little further progress was made until the Renaissance. Nicolaus Copernicus showed that instead of the Sun daily revolving around Earth, Earth travels around the Sun on an annual basis. Later, the advent of the telescope enabled Galileo Galilei to discover the Moon to be a world possessing a vivid landscape.

Next Johannes Kepler realised the orbits of the planets and their satellites aren't circles but ellipses. This enabled him to develop three laws of planetary motion.

Along with an apple, the Moon played a key role in Isaac Newton's discovery of gravitation. In concert with three laws of motion that define the nature of physical systems, gravitation explained Kepler's empirical laws.

It was hoped that knowledge of the motion of the Moon across the sky and the character of its surface would provide a way to determine the longitude of a ship at sea. New telescopes enabled people to map the Moon with ever increasing sophistication, and photographic atlases were produced in the late 19th century. But studies revealed irregularities in the motion of the Moon that made using it as a clock impracticable. On the other hand, by then the 'longitude problem' had been solved by the invention of the marine chronometer.

However, the Moon had also become an object of fascination in its own right. In addition to mapping the Moon, people gave thought to the process that had created its distinctive surface features. Arguments raged about whether the process involved volcanism or the impact of celestial objects such as wayward asteroids.

As astronomers investigated ever finer details of the lunar surface, the Moon attracted the interest of geologists, whose insights revealed the history of that world in ways never imagined by astronomers. In general, however, astronomers resented this intrusion into their bailiwick.

In the mid-20th century we developed the means to send automated probes to investigate the Moon by going into orbit around it and by landing on its surface.

As astronauts prepared to make the first human landing on the Moon, scientists continued to argue about what they would find there.

The 'ground truth' provided by the samples returned by the early landings supported some hypotheses and shot down others.

Later missions were field trips designed to understand specific aspects of the nature of the lunar surface and its history.

There was then a considerable pause as the results from this incredible early period of lunar astronautics were assessed and consolidated. One result was the rejection of all the theories advanced prior to the Space Age to explain the origin of the Moon. In the ensuing decade a new hypothesis was developed; not, of course, that everyone agrees with it!

In the 1990s the first of a new wave of probes used sophisticated sensors to survey the composition of the lunar surface, to chart its topography, and to map gravitational irregularities for insight into the internal structure. In this century, other nations have started to send probes, turning the scientific investigation of the Moon into a truly international venture.

One major discovery was that there is water in polar craters whose floors are in permanent shadow. When we return to the Moon, we will very likely establish a base of operations on a nearby patch of high ground where the Sun always shines to provide the power to mine the resources which will enable us to venture to destinations beyond.

OPPOSITE The launch of Apollo 11 on 16 July 1969. *(NASA)*

Chapter One

Classical times

This chapter reviews how Greek and Egyptian philosophers studied the motions of the Sun, the Moon, and the planets, resulting in the development of a system that was dominant for over a millennium.

OPPOSITE *The School of Athens* was painted by Raffaello Sanzio in 1509. It depicts the greatest thinkers of classical antiquity. This portion of the fresco features Plato (left) and Aristotle (right). The original is currently in the Vatican.

In its 'golden age' spanning 600 to 300 BC, Greece was the centre of the arts and natural philosophy. Visitors from afar would bring new knowledge, which the Greeks assimilated and then diffused across their domain.

The first philosopher in this tradition was Thales of Miletus, a city of Ionia, which was a state on the western coast of Asia Minor (now Anatolia). As one of the Seven Sages of Greece, Thales is considered to be the 'father of science' because he tried to explain natural phenomena without reference to mythology, as had been the previous way. At the end of the 7th century BC Thales introduced his fellow philosophers to the achievements of Egyptian astronomy.

Thales realised that the Moon shines by reflected sunlight and it becomes invisible when its path takes it close to the Sun.

Pythagoras was born in 570 BC on Samos, a Greek island off Ionia that became a crossroads between Asia, Africa, and Europe. In his youth he visited the elderly Thales. In addition to famously studying geometry, he realised that the Moon is a sphere and reasoned by analogy that Earth must also be spherical. He also asserted that Earth must be at the centre of the universe and that there is a fundamental boundary whereby the Moon and everything 'above' is 'perfect', whereas Earth is subject to change and therefore to decay. When critics asked about the markings on the face of the Moon, he suggested it was a mirror and its markings were really a distorted reflection of Earth.

Plato, Eudoxus and Aristotle

In the 5th century BC, the Athenian scholar Socrates devised the 'dialectic' method of inquiry whereby a problem is broken down into a series of questions, the answering of which leads to the eventual resolution of the original question. Around 387 BC, Plato, a student of Socrates, founded the Academy in Athens as an institution of higher learning. With his mentor and his own student, Aristotle, Plato developed the basis of Western philosophy and science.

In particular, Plato dismissed the act of observing the motions of celestial objects as a task that was beneath a true philosopher, who ought to be able to reason from first principles.

Plato believed that celestial motions must be both uniform and orderly, and therefore be amenable to description by mathematics. In particular he reasoned that the Moon, Sun, and stars resided on concentric spheres that were centred on Earth.

Eudoxus was born in 409 BC in Knidos, an ancient Spartan city located in southwestern Asia Minor. On his first visit to Athens he became a student of Plato. After studying in Heliopolis in Egypt, he established a school of philosophy in Cyzicus on the shore of the Sea of Marmara. At the urging of his mentor, Eudoxus accepted the challenge of explaining why some planets trace out irregular 'loops' in the sky. His devised a scheme which combined spheres and uniform circular motions. The key point, which delighted Plato, was that the paths of the planets were indeed open to geometric description.

By showing that his scheme could account for the types of motion which were known to occur, Eudoxus had, in modern parlance, provided a 'proof of concept'. Working out the details would not be possible until someone of a more practical nature compiled measurements.

Aristotle, the most famous of Plato's students, was born in 384 BC into a wealthy family in the Macedonian city of Stagira. After his mentor died, Aristotle remained in Athens and established his own school of philosophy called the Lyceum and spent the remainder of his life writing accounts of the works of his predecessors, thus producing a comprehensive review of the state of human knowledge at that time. In fact, many of the texts that Aristotle cited have not survived and hence we know of them only by way of his assessments.

Like Pythagoras, Aristotle believed that the celestial realm must be 'perfect', therefore the Sun, Moon, planets, and stars must all revolve around Earth in circles because, as he asserted, 'All that is eternal is circular.'

Aristotle also reasoned that the concentric spheres which Plato associated with the Moon, Sun, and planets were real structures made of a solid transparent material, and that they rotated around Earth at different rates in a frictionless manner. The stars were on another sphere that enclosed the system and formed the perimeter of the universe.

After the time of Aristotle, the centre of Greek

scientific thought moved to Alexandria. Founded by Alexander the Great in 332 BC, this great city was the capital of Egypt during the reigns of a succession of Macedonian rulers, all of whom took the name Ptolemy. As patrons of learning, they established a museum which incorporated an astronomical observatory and a library. Since it was the law that any ship that sailed into port must donate copies of books to the library, this collection soon became the best in the world.

Apollonius, Hipparchus and Ptolemy

Astronomers had noted that the path of the Moon across the sky was inclined to Earth's equator and that, although they could predict its position in general terms, it was often ahead of, and at other times in trail of its ideal position.

In the 3rd century BC, Apollonius of Perga in Asia Minor offered an explanation for this variation in the rate at which the Moon travelled across the sky. In this geometrical scheme, the Moon was actually moving on a small circle whose central point was tracing out a large circle around Earth; the small circle was termed the *epicycle* and the large circle was the *deferent*. This also accounted for the observation that the size of the Moon in the sky appeared to vary in a cyclical manner. Because the scheme involved only circles it had the virtue of aesthetic purity.

Hipparchus lived in the 2nd century BC. In contrast to Plato, whose interest in astronomy was simply an exercise in geometry, Hipparchus recognised the value of making observations and measurements. He may have visited Alexandria, but it was on the island of Rhodes in the Aegean that he constructed an observatory and carried out most of his work. He went to great lengths to ensure that his celestial observations exploited the precision of his instruments. In the process, he compiled a catalogue of over 1,000 stars, noting both the position and brightness of each.

On the basis of his measurements of the Moon's path, Hipparchus concluded that Apollonius's scheme involving epicycles and deferents was satisfactory.

Unfortunately Hipparchus was unable to investigate the motions of the planets in detail because there were so few reliable archived measurements of their positions. He therefore compiled a record of his own observations in the hope of enabling his successors to conduct such an analysis.

Eratosthenes was born in 276 BC in Cyrene, the city that gave the name Cyrenaica to what is now the eastern portion of Libya. He became the chief librarian in Alexandria and while there, made an important measurement. The Sun moves against the stars, completing a revolution of the sky in one year. Its celestial path, known as the *ecliptic*, was clearly inclined to the equator. Eratosthenes was the first to measure this obliquity as 23.5°.

The ecliptic and equatorial planes intersect at two points, known as *nodes*. The point where the Sun crosses the equator northbound is called the *vernal equinox* and that where it crosses southbound is the *autumnal equinox*.

By comparing his own measurements of the positions of stars with those of his predecessors

BELOW The scheme by which a body travels in a small circle called an epicycle whose centre travels around a larger circle called a deferent. *(Woods)*

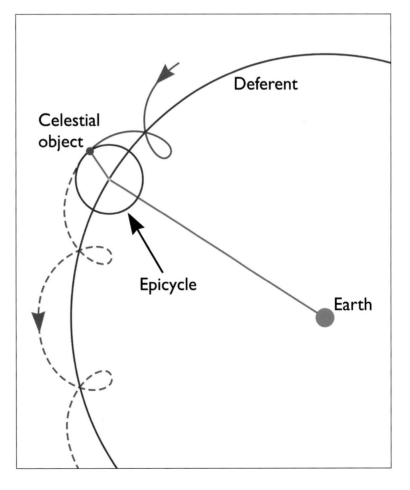

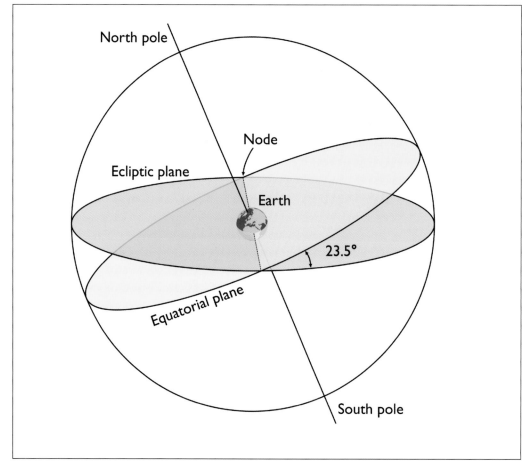

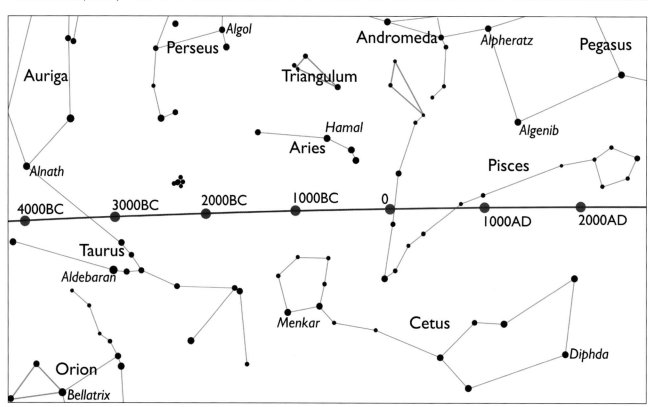

150 years earlier, Hipparchus noted a systematic drift that implied the ecliptic plane to be rotating with respect to the equator, causing the precession of the equinoxes over a period of tens of thousands of years.

For the next several centuries the history of astronomy is essentially a blank.

The last great name in Greek astronomy was Claudius of Ptolemais, the capital city of Upper Egypt. Nothing is known of his life except that he was born around AD 85 and lived most of his life in Alexandria, which was then controlled by Rome.

Ptolemy (as he is known in English) used his own observations and the resources of the Alexandrian library to refine the studies of Hipparchus, and then wrote a book in which he explained the works of earlier astronomers. The original Greek manuscript was later translated into Arabic. When subsequently translated into Latin, the Arabic title of *Al Magisti* (*The Greatest Treatise*) resulted in it becoming known in English as the *Almagest*. It is of tremendous historical significance, since without it we would know very little of Greek astronomy.

Hipparchus had explained the motion of the Moon remarkably well using Apollonius's scheme of epicycles and deferents, so Ptolemy decided he would tackle the issue of planetary motions in the same terms. He thus reinterpreted the uniform circular motions envisaged by Eudoxus, added some ideas of his own, and then used the observations which Hipparchus had compiled, augmented with his own, to work out the details.

In order of increasing distance from the centrally located and immobile Earth, Ptolemy placed the Moon, Mercury, Venus, the Sun, Mars, Jupiter, and Saturn, with the stars on a sphere beyond. This *Ptolemaic system* of planetary motions was the first to provide real predictive power.

Later the Church of Rome adopted Aristotle's philosophical teachings, and the Ptolemaic system of planetary motions became its official doctrine.

The Moon and the calendar

One of the chief tasks of an astronomer in ancient times was to regulate the calendar. The primary reference was the cycle of day and night defined by the Sun. Next was the progress of the Moon through its illumination cycle – sometimes not being visible at all, then *waxing* through the crescent shape to half of a full disc, then to the three-quarter 'gibbous' shape and hence to a full disc, and thereafter *waning* in reverse through to the opposite crescent prior to vanishing again. The changing shape of the lunar disc in this cycle is known as its *phase*. The complete cycle, known as a *lunation*, takes 29.53 days.

It was the Athenian statesman Solon who, in the 6th century BC, coined the terms 'old' and 'new' for the waning Moon that vanishes into the predawn glow and the one that appears after sunset several days later. This gave rise to the concept of a month. The early calendars were therefore based on the Moon.

In addition, astronomers arranged the stars into patterns called constellations and observed that these changed with the seasons, giving rise to the concept of a year.

A lunar calendar was excellent for a nomadic tribe which required only a daily period and the ability to refer to events as having occurred 'many moons ago'.

An agricultural society, however, must take account of the seasons in order to know when to plant and harvest crops.

The early Egyptian calendar was based on the Moon but they used Sirius, the brightest star in the sky, to predict the impending annual Nile flood. They knew the year didn't divide precisely into months and declared the 'extra' days at the end of the year to be a festival that was 'outside' the calendar.

A year with 12 months of 30 days gave a

BELOW Diagram to show how the Egyptians incorporated the rising and setting of Sirius into their calendar.
(Woods)

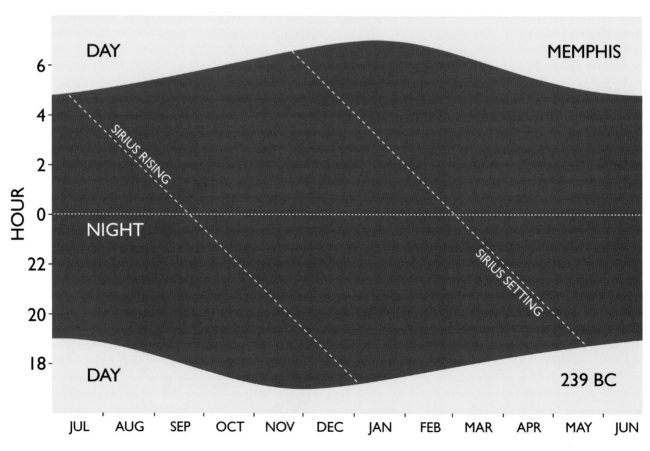

total of 360 days, a mathematically attractive number, but it made both the month and the seasons drift progressively. A scheme in which a year had 6 months of 30 days and 6 months of 29 days provided a month of 29.5 days but the year of 354 days made the seasons drift even more rapidly.

In an attmept to fix this, around 594 BC Solon introduced the refinement of adding one 30-day month to each alternating year. That is, there were $13 \times 30 + 12 \times 29 = 738$ days during every 2-year period. Over a total of 25 months this provided an average of 369 days per year and 29.5 days per month. It was another excellent match to the Moon but the seasons were still drifting, this time in the opposite sense.

A further refinement in the 5th century BC was an 8-year cycle involving 5 years in which there were 6 months of 30 days and another 6 months of 29 days, plus 3 years with 7 months of 30 days and 6 months of 29 days. The total of 2,922 days not only provided an average month of 29.5 days but also an average year of 365.25 days to stabilise the seasons.

The Babylonians developed a calendar which was rich with astrological significance. Their year had 12 months, each of which began with the first sighting of a 'new' crescent Moon. Their months were therefore each of precisely one lunar cycle. To accommodate the fact that this did not divide into the year, they added an extra 13th month with a duration that would ensure that any given month would always occur at the same time of the year. They also introduced the concept of a week by making the first day of a month the first day of a new week.

The alert Babylonians also realised that 235 lunar cycles were almost precisely 19 years; the difference being less than one day. This greatly assisted in regulating their calendar in the long term.

On learning of this pattern circa 460 BC, the Athenian astronomer Meton proposed a calendar that had 12 years of 12 months and 7 years of 13 months, such that 125 months had 30 days and 110 months had 29 days. This provided a total of 6,940 days over a single 19-year cycle, giving an average of 365.26 days per year and 29.5 days per month.

This Metonic cycle enabled the lunar and solar calendars to resynchronise every 19 years. Although it wasn't adopted as a civic calendar, it was an accurate means of determining when to apply a correction to the calendar that *was* in use.

Interestingly, the Metonic cycle still serves a purpose today, being used to calculate the date of Easter, which is related to the position of the Moon.

The early Roman calendar had a 354-day year of 12 months, with an extra month of 22/23 days added every 2 years. This calendar began in 754 BC, the traditional date for the founding of Rome. When it had drifted so far as to have become impractical, Emperor Julius Caesar commissioned Sosigenes of Alexandria to create a better calendar. The catch-up was achieved by adding 3 additional months to 46 BC, making that year last for 445 days. The Julian calendar that began the next year employed a 4-year cycle in which there were 3 years of 365 days and a *leap year* of 366 days to give the desired average of 365.25 days.

Sosigenes didn't attempt to regulate the calendar with the lunar cycle. Any month which had more than 29.5 days would inevitably span more than one lunar cycle, thus containing either two 'new' or 'full' Moons. The second 'full' Moon in any month is called a Blue Moon, a term that is nowadays used to imply an event which doesn't occur very often. Given the dominance of the Roman Empire, the Julian calendar was soon adopted throughout Europe.

The difference between the average length of a Julian year and a true year was only 1 day in a span of 128 years. Nevertheless the drift became inconvenient. In 1582 Pope Gregory XIII ordered an adjustment in which the date was immediately advanced by 10 days to catch up, and the leap-year scheme was refined in order to slow the rate of drift.

This Gregorian calendar is in common use today but some countries refused to make the change. Thus the 'October Revolution' in Russia actually took place in early November. Furthermore, the Islamic and the Jewish calendars are based on the Moon, with their months beginning with the first sighting of the waxing crescent; hence each day starts at sunset.

Chapter Two

The Renaissance

For over a millennium our view of the natural world was dominated by the teachings of Aristotle, but this old order was displaced during the Renaissance by a renewal of critical thinking, a resumption of measuring the positions of celestial objects, and the invention of the telescope.

OPPOSITE A 19th century painting by Joseph-Nicolas Robert-Fleury depicting Galileo, at 70 years of age, defending himself in front of the Inquisition of Rome on 22 June 1633. The presiding cardinal confronts him and the judges sit at the bench beyond.

In 1330 the Italian scholar Francesco Petrarca coined the term *Dark Ages* for the centuries of cultural decline that afflicted Europe after the fall of Rome in the 5th century. During this time, many of the works of classical Greece and Rome were available only in Arabic translation.

Turmoil in Constantinople in the early 13th century prompted many Byzantine artists and academics to flee to Italy, and particularly to Florence, which was the capital of the Grand Duchy of Tuscany in the north, where their reintroduction of classical Greek styles paved the way for what would become known as the Renaissance (Rebirth). European philosophers soon developed a renewed interest in astronomy and set about translating Ptolemy's *Almagest* from Arabic into Latin.

While a student at the University of Vienna in Austria, Johannes Müller of Königsberg in the Duchy of Franconia, now Bavaria, developed an interest in mathematics and astronomy. In 1461, aged 25 and now a member of the clergy, Müller was invited to Rome to inspect an original Greek copy of the *Almagest* which had been obtained after the fall of Constantinople to the Turks in 1453. Mistakes in the Arabic version had been carried over when that was translated into Latin, so the safest way forward was to translate the Greek directly into Latin.

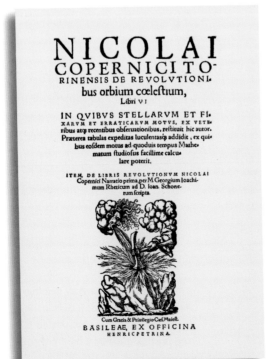

RIGHT The title page of Nicolaus Copernicus's book *De Revolutionibus Orbium Coelestium*.

Nicolaus Copernicus

Nicolaus Copernicus was born in 1473, the youngest of four children. His father was a merchant who in 1462 had relocated from the Polish capital of Cracow to the small town of Toruń on the Vistula, which was then under the tenuous control of the King of Poland. On the death of his father in 1483 Copernicus went to live with his uncle, a cleric in the cathedral city of Frauenburg on the Baltic. At the age of 17 he enrolled at the University of Cracow, where he developed an interest in astronomy, the main texts for which were the works of Johannes Müller.

Leaving the university after three years without graduating, Copernicus first failed a candidature for an ecclesiastic career and then spent a decade in Italy, notionally studying medicine and law but in reality furthering his astronomy. As was typical of astronomers of that time, his interest in the subject was theoretical; he had little desire to make observations.

On returning to his uncle in 1506 Copernicus spent several years writing a book in which he asserted that Earth was in motion around the Sun. He refrained from publishing his work because he didn't wish to endure the controversy he knew would be stirred up.

By 1530 his book was in as refined a form as he could wish, but still he resisted publishing it. Instead, he privately circulated his conclusions in a brief summary paper. This document prompted the Pope in 1533 to invite a Vatican astronomer to deliver a lecture on the radical hypothesis. To facilitate this, Copernicus was asked to supply additional information.

In 1539 Copernicus finally decided to arrange for his book to be published, but the process was protracted. He suffered a stroke in December 1542 and died six months later, shortly before his tome *De Revolutionibus Orbium Coelestium* (*On the Revolutions of the Celestial Spheres*) was finally issued.

So what was all the fuss about?

Early in his studies, Copernicus had become dissatisfied with the Ptolemaic system. During his research into the works of earlier philosophers he found references to different beliefs. In particular, Philolaus, a Greek who lived in southern Italy a century after Pythagoras, was the first to suggest that Earth not only turned on

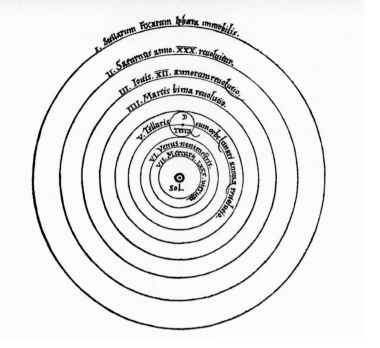

its axis but was also in motion. But the scheme Philolaus offered was rather fanciful. And whilst Aristotle had conceded that the apparent daily motion of the stars *could* be explained by a motion of either the stars or of Earth, he *rejected* the latter explanation. By his insistence that Earth resided at the centre of the universe he ruled out the possibility that Earth could be in motion.

Copernicus also found older Egyptian arguments asserting that because Mercury and Venus never strayed far from the Sun in the sky they must actually travel around the Sun; as opposed to the other planets, whose paths took them completely around the sky. But the Aristotelian assertion that only Earth was the centre of motion had prevailed.

Inspired by these early writings, Copernicus set out to study the situation for himself, and he concluded that the apparent motions of celestial bodies are a combination of the consequence of Earth rotating daily, the Moon revolving around Earth on a monthly basis, Earth revolving around the Sun on an annual basis, and the planets doing likewise in their own individual periods. In order of increasing distance from the Sun, there was Mercury, Venus, Earth and its attendant Moon, Mars, Jupiter, and Saturn; beyond that, as in the Ptolemaic system, were the stars.

Because Helios was the name of the Greek god of the Sun, Copernicus named his scheme the *heliocentric hypothesis*. Its chief merit was that whereas previous philosophers had merely speculated without evidence, Copernicus had worked it out with mathematical rigour. On the downside, the book was unintelligible to the lay reader.

Heliocentrism was superior to the teachings of Aristotle because it was a simpler explanation of celestial motions. But like Ptolemy, Copernicus believed Aristotle's view that celestial motions must be circular and so he used epicycles to explain the observed irregularities.

Galileo Galilei

Galileo was born on 15 February 1564 as the first son of Vincenzio Galilei, a musician who had relocated his family from his native Florence to Pisa on the coast of Tuscany, where he became a merchant. In the 12th century, Pisa was one of Italy's leading mercantile cities, but by Galileo's time it was subsidiary to Florence.

Aged 17, Galileo attended the University of Pisa to study medicine to please his father, but he left after three years due to financial hardship without having gained his degree. Rejoining his family, which had returned to Florence, he pursued his interests in physics and astronomy privately for the next four years, learning as much as his contemporaries; indeed, in some ways he became better informed.

At the age of 25 he gained a professorship at his alma mater to teach mathematics. In this role, he carried out experiments into how objects fall. Ever since Aristotle in classical times, it had been presumed that the time taken by a body to fall a given distance was inversely proportional to its weight. That is, a body with twice the weight of another would fall twice as fast and therefore cover the same distance

ABOVE LEFT
Nicolaus Copernicus.

ABOVE **The heliocentric system as depicted by Copernicus in his book.**

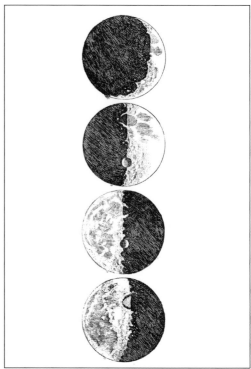

ABOVE In Galileo's time, Italy consisted of a number of city states with different affiliations with the Vatican. *(Harland)*

ABOVE RIGHT In his *Sidereus Nuncius*, Galileo depicted the Moon under several illumination phases, indicating that it was a world in its own right with a highly distinctive landscape.

in half the time. Galileo, as was his way, set out to *test* this. The story is that he dropped two stone balls, one weighing ten times more than the other, off the Leaning Tower. The results showed that the weights hit the ground essentially simultaneously. So much for dogma!

Galileo saw that a heavier weight did indeed fall *slightly* faster, but he believed the lighter one had been retarded by the resistance of passing through air and suggested this difference would be eliminated if the experiment could be done in an airless chamber – a vacuum, however, was something that Aristotle had deemed impossible to create.

Galileo also realised that a ball that was set rolling on a horizontal surface in the absence of air resistance would come to rest only due to friction with the surface. He then conjectured that if all the retardations could be eliminated, a ball would roll *forever* in a straight path. This was a radical thought, because Aristotle had insisted that in order for an object to remain in uniform motion a force must be *continuously* applied. Yet Galileo had concluded that, in ideal conditions, *no* such force was required.

A further advance was measuring the rate at which the speed of a falling body increased – i.e. its acceleration. It was not practicable to attempt this experiment by dropping balls

vertically, so Galileo rolled them down a sloping track to slow the rate of vertical travel. His knowledge of mathematics told him that this was a valid test of how an object fell. To accurately measure very brief intervals of time, he used a 'water clock', a device invented by the Egyptians which could provide a relative measure of duration. The results showed that a body is uniformly accelerated when, starting from rest, it acquires equal increments of speed during equal intervals of time.

Instead of reasoning on the basis of what accepted wisdom said ought to be true, Galileo had measured what actually occurred and expressed the results mathematically. It was a considerable advance over the Aristotelian approach.

As Galileo's research continued to contradict Aristotle, his academic colleagues became angry, and after three years he was driven from his post and returned once again to Florence.

The following year, 1592, Galileo received a professorship of mathematics at the University of Padua, which was the leading Italian university. The Venetian Republic in the east had different politics from Pisa, and his new colleagues were receptive to his ideas. For the next 18 years he was extremely productive. Shortly after settling in Padua, Galileo became an eager convert to the heliocentric hypothesis of Nicolaus Copernicus.

During a visit to Venice in July 1609 Galileo heard, via a letter written to one of his friends by a French nobleman, that a Dutchman had invented a device that used lenses to enable an observer to see distant objects as if they were much closer. Galileo promptly made his own telescope. When he demonstrated it to the Senate a month later, showing that a merchant on a tall building could identify an inbound ship several hours earlier than was otherwise possible (knowledge which would be commercially valuable in a city of merchants) he was rewarded not only with an increase in his salary but also by having his professorship made permanent.

On turning his telescope to the heavens, Galileo immediately discovered further evidence to contradict Aristotle. Starting on 30 November he observed the Moon several times over 18 nights and drew a series of sketches which showed that although the outer curve was sharply defined, the inner edge that varied from night to night as the illumination phase changed was irregular.

Aristotle had taught that the Moon, being in the celestial realm, ought to be a perfect sphere and it had been argued the features on its face were a reflection of Earth. Clearly this wasn't the case. Galileo realised the Moon was a world in its own right.

In the 6th century BC Pythagoras had realised that the Moon is a sphere that doesn't shine by its own light but reflects sunlight, with

the phases being determined by how much of the illuminated hemisphere is visible.

Leonardo da Vinci, an Italian contemporary of Copernicus, was the first to correctly explain the faint illumination that appears over the majority of the lunar disc at the time of a thin crescent. He pointed out that when the Moon is almost 'new', that hemisphere of Earth which is illuminated by the Sun reflects sufficient light onto the dark part of the lunar disc to render it visible, producing the spectacle poetically known as 'the old Moon in the young Moon's arms'.

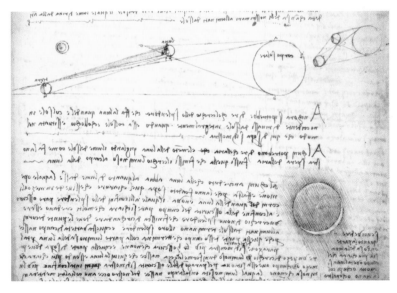

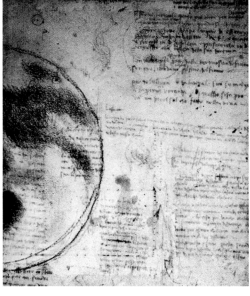

In 1505 da Vinci, who possessed exceptional eyesight, drew an impression of the face of the Moon. He reasoned that the brighter portion was water, the dark areas were land, and there were clouds in the lunar atmosphere.

In the 1590s William Gilbert, who was one of Queen Elizabeth I of England's physicians, made a study of the Moon and in 1603 he sketched a naked-eye map. It was included in a posthumous summary of his work *De Mundo Nostro Sublunari Philosophia Nova*

(New Philosophy about our Sublunary World) published half a century later. Although his map was rudimentary, he gained the distinction of being the first to assign names to prominent features. Like da Vinci, he believed the light patches were water.

With the benefit of a telescope, Galileo found that both men had been misled; the *bright* areas were mountainous and the vast dark areas were more likely to be seas (although he didn't assert they were). However, Galileo saw no evidence of clouds. He was impressed by the shadows which revealed no detail. He made use of the shadows cast by mountains to estimate their heights. This was straightforward in theory but difficult in practice and later observers using better instruments found his heights to be exaggerated, but the significant point is that he had gained some impression of the scale of lunar topography.

When Jupiter was well placed for viewing in January 1610, Galileo discovered it to be accompanied by four tiny points of light that lay in a straight line and moved to and fro in a manner that suggested they were travelling around the planet. Jupiter possessed a quartet of moons. Contrary to the Aristotelian view, Earth was not the only centre of motion in the celestial realm.

On 12 March Galileo published his various observations in a pamphlet entitled *Sidereus Nuncius (Sidereal Messenger)*.

The principle of the telescope seems to have been identified many times. In the 13th

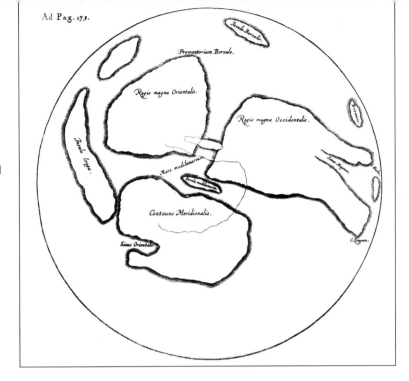

century it was found that a ray of light was bent by passing it through a glass lens. At that time Roger Bacon in England wrote of using a combination of lenses to make distant objects appear closer. Leonardo da Vinci speculated in 1490 about whether lenses in combination could produce an enlarged view of a distant object. He conducted tests in 1504 and by 1510 had determined the principle of the telescope. Several years later he described how a concave mirror could create a magnified image.

In a book entitled *Pantometria* published in 1570, Thomas Digges in England explained how his father, Leonard Digges, having invented the theodolite in his profession as a surveyor, also created a 'proportional glass' which enabled him to view distant objects and people. This concept was independently discussed by Giovanni Battista della Porta in Italy in 1558.

But none of these ventures had any lasting impact. The breakthrough was achieved by Hans Lippershey. Born in Germany in 1570 he became a citizen of Flemish Zeeland in 1602 and worked as a spectacle-maker in Middleberg. Upon assembling a telescope he applied to the Hague for a patent in October 1608 (it wasn't granted) and started selling them.

In fact, Galileo wasn't the first to turn a telescope to the heavens. After graduating in mathematics from Oxford in England, Thomas Harriot tutored Sir Walter Raleigh on navigational issues and several times sailed with him. By 1603 he was comfortably established in London, pursuing his interest in optics. A comet in 1607 attracted him to astronomy, and on 26 July 1609, having recently obtained a crude telescope from Holland, he aimed it at the Moon.

As Galileo had a leaning towards painting, he had made representational 'portraits' of the Moon on which few features are recognisable.

ABOVE When William Gilbert sketched the face of the Moon in 1603 he assigned some names.

FAR LEFT The title page of Galileo's 1610 pamphlet *Sidereus Nuncius*. (The Houghton Library of Harvard University)

LEFT Portrait reputedly of Thomas Harriot.

What is more, he never drew a full disc to consolidate his observations. After 1609 Galileo paid scant attention to the Moon.

Harriot continued his lunar observations for several years and sought fidelity when sketching it. In 1611 he compiled a whole-disc map whose features are clearly recognisable. But he never published this, and it wasn't unearthed in his papers until long after his death. Where Galileo had excelled was in public relations, which is why we all know his name.

In September 1610, Galileo saw that Saturn had what seemed to be two large companions whose behaviour was odd because they did *not* move. Another revelation in that month was that the Sun is a sphere, rotating on its axis with its surface marred by dark spots; imperfections which were contrary to Aristotelian teaching.

But Galileo's most significant discovery came in October when he found that Venus undergoes phases like those of the Moon. The way the phases correlated with the variation in the apparent diameter of the planet *proved* that, contrary to the Ptolemaic system, it really does travel around the Sun as the ancient Egyptians had surmised and as Copernicus had insisted that it must.

There was now a clear distinction between the Vatican's belief in the Aristotelian concept of the celestial realm and its reality as revealed by telescopic observation.

Although Galileo had life-tenure at Padua, in the autumn of 1610 he returned to the University of Pisa as head of mathematics with sponsorship from the Tuscan court in Florence, the latter doubtless prompted by the fact that he had dedicated his pamphlet to his former pupil, Cosimo de Medici who had become the Grand Duke of Tuscany. His intention was to teach heliocentrism, which was risky. If he had remained in Padua then the vigorously proclaimed independence of the Venetian Republic would have insulated him from the Church of Rome, but Tuscany was more closely aligned with the Vatican.

The Church wasn't overly concerned about Galileo challenging Aristotle with 'imperfections' on the Moon, but demoting Earth from its central role in the universe was much more serious.

Under pressure from Rome, the University of Pisa directed Galileo not to teach his contentious theories but, as was his way, he refused to comply.

In December 1614 Father Thomas Caccini delivered a sermon in Florence which indirectly censured Galileo by denouncing those elements of mathematics that were inconsistent with the Bible. The cleric then went to Rome and informed the Inquisition. On hearing of this, Galileo decided to go to Rome to explain his position. When he did so in December 1615 Galileo was informed that the Copernican theory was incorrect and he must not defend it. He would be free to pursue his researches, so long as he refrained from publicly contradicting accepted doctrines.

In the summer of 1616 Galileo returned to Florence, satisfied that he had gotten off lightly.

Indeed, he had. One of his contemporaries, Giordano Bruno, an Italian Dominican friar, had asserted firstly that there couldn't be a celestial body at the centre of the universe because the universe, being infinite, has no centre, and also that the Sun wasn't special because the stars were distant suns. The Inquisition had judged Bruno guilty of heresy and, when he refused to recant, had ordered him burned at the stake in 1600.

In 1624, a year after the election of a new Pope, Galileo went to Rome to try to have

BELOW A map of the Moon by Thomas Harriot, consolidating many of his observations. *(Petworth House Collection, HMC 241/9, ff. 26–30. Courtesy of Lord Egremont and Leconfield)*

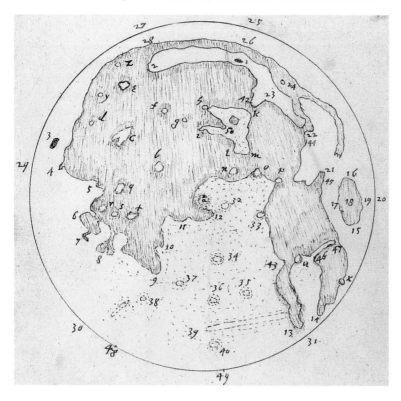

FAR LEFT **A portrait of Giordano Bruno.**

LEFT **A portrait of Galileo in 1624 by Ottavio Leoni.**

the restriction on his teaching lifted, but this wasn't granted.

Undeterred, at the age of 60 Galileo set about writing a book entitled *Dialogue Concerning the Two Great World Systems, the Copernican and the Ptolemaic*. All books in Catholic territory needed to pass a Vatican censor prior to publication. The book by Copernicus was on the prohibited list, but Galileo's manuscript was approved with minor revisions and published in 1632.

Galileo had formulated his text in the form of an imagined discussion between someone who believed wholeheartedly in Aristotle, a man who spoke as Galileo himself believed, and a layman who was open to persuasion. However, despite the book considering the issues in an impartial and noncommittal manner in order not to openly contradict Church doctrines, the Pope suspected that he was the stubborn Aristotelian, and so the book was withdrawn.

On 22 June 1633 the Inquisition found Galileo guilty of heresy for expressing his belief in heliocentrism. After offering an apology, he was allowed to return to Florence to live out his life under house arrest. In this state of isolation he wrote a new treatise and in 1638 had it published in Holland which, being Protestant, was safely beyond the reach of the Inquisition. Having achieved the age of 78, he died of a fever in 1642.

Fully a century after Copernicus expressed his opinion that the Sun, not Earth, was at the centre of the universe, the Vatican continued to denounce this as heresy despite the truth being evident to anyone who cared to turn a telescope to the heavens.

In 1822 the Vatican decided the 'publication of works treating of the motion of the Earth and the stability of the Sun, in accordance with the opinion of modern astronomers, is permitted'. It wasn't an admission that its doctrines were incorrect, merely a concession that Earth *might* revolve around the Sun.

Another 13 years elapsed before the writings of Copernicus and Galileo were removed from the list of prohibited publications, known as the *Index Librorum Prohibitorum*.

It was not until 1992 that the Vatican admitted Earth isn't stationary in the heavens.

Finally, in 2000, Pope John Paul II apologised for the way in which Galileo had been treated.

BELOW **The title page of Galileo's *Dialogue Concerning the Two Great World Systems, the Copernican and the Ptolemaic*.**

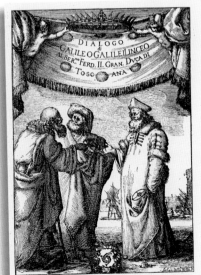

Chapter Three

Celestial order

A renewal of observational astronomy by Tycho Brahe resulted in Johannes Kepler deriving mathematical relationships for the motions of the Moon, Earth, and the planets, and subsequently to Isaac Newton using the Moon to explain Kepler's empirical laws in terms of the universal force of gravity.

OPPOSITE Isaac Newton, sitting on the ground beneath a tree musing upon his research, has an epiphany as an apple falls nearby. *(The Florida Center for Instructional Technology)*

After Hipparchus compiled a catalogue of celestial measurements during the 2nd century BC, it was many centuries before astronomers resumed such work.

Tycho Brahe

Tycho Brahe was born in 1546 at Knudstrup in the Danish province of Scania, now the southern tip of Sweden. As the eldest child of a nobleman, he was adopted in his infancy by an uncle and raised on a country estate.

Brahe enrolled at Copenhagen University at the age of 13 with a view to entering politics, but in 1560 a solar eclipse spurred an interest in mathematics and astronomy. On relocating to the University of Leipzig several years later to focus upon his astronomical studies he became frustrated with the inaccuracy

RIGHT Uraniborg was a grand facility, with its outer walls aligned to the four points of the compass. From *Astronomiæ Instauratæ Mechanica.*

of the positions predicted for Jupiter and Saturn. At this time, Brahe bought a number of astronomical instruments and set out to identify their intrinsic accuracies, and determine whether those errors could be minimised by using systematic observing techniques.

Beginning in 1566 Brahe travelled widely in Germany visiting other astronomers, for one of whom he constructed a quadrant (quarter-circle) instrument which had a radius of 5.7m with the scale upon its rim divided into single minutes of arc.

When a 'new star' appeared in 1572 in the constellation of Cassiopeia, which at its brightest was visible in daylight, Brahe monitored its slow fading from view over a period of 16 months and then published his observations.

In 1576 Frederick II of Denmark, who was a patron of the arts and sciences, provided Brahe with annual endowments to enable him to carry out astronomical work in the most effective way. This included use of the low-lying island of Hven in the strait between Scania and Denmark for an observatory, accommodation for himself plus his technical assistants, a library, and the workshops in which to create instruments of his own design. Brahe named this establishment Uraniborg (Castle of the Heavens) and it soon became the finest astronomical observatory to that time. He described his tools in the 1598 treatise *Astronomiæ Instauratæ Mechanica* (*Instruments for the Restoration of Astronomy*).

Brahe decided to adopt a policy of observing as many celestial objects as possible, as often as possible, in order to accumulate an essentially continuous record over as prolonged

RIGHT An engraving of Tycho Brahe in his Uraniborg observatory on the island of Hven. He is pointing to a slot in the wall through which an assistant sights the large quadrant. It is a hand-coloured engraving from *Astronomiæ Instauratæ Mechanica.*

a period as possible. It demanded remarkable stamina and consumed the remainder of his life.

As Brahe's understanding of the sources of error in making positional celestial observations increased during the years, the accuracy of his measurements improved from an initial typical error of several minutes of arc to better than a single minute of arc. The result was a series of observations that greatly transcended both the scope and the accuracy of his predecessors.

When Frederick II died in 1588 the throne passed to his 11-year-old son. A committee of noblemen ran the government until Christian IV came of age in 1596. Support for Brahe's work was withdrawn one year later and the year after that he packed his possessions and moved to Hamburg. On accepting the patronage of the Habsburg Emperor Rudolph II in the summer of 1599, he relocated to Prague. In 1600 he gained the services of Johannes Kepler as an assistant.

BELOW A painting of Johannes Kepler in 1610 by an unknown artist.

Johannes Kepler

In 1571 Johannes Kepler became the fourth child of an impoverished family in Weil der Stadt, Württemberg. He attended the monastic school in Adelberg at the age of 13, and progressed to college in Maulbronn. On graduating in 1588 he went to the University of Tübingen, which was at that time one of the great centres of Protestant theology. There one of his mathematics tutors introduced him to the works of Copernicus as a counterpoint to the traditional formal lectures.

In 1594 Kepler was hired as a mathematics teacher at the Lutheran high school in Gratz, Austria, where he made use of his knowledge of astronomy to supplement his income by making astrological predictions. In a wave of religious intolerance, all Protestant clerics and teachers were ordered to leave Gratz. Kepler first fled to Hungary and then in 1600 was recruited by Brahe to assist with his astronomical research.

In 1602, shortly after Brahe's death, Kepler was appointed to succeed him as mathematician to Emperor Rudolph II, albeit at a much reduced remuneration. Having no interest in continuing the observational work, Kepler devoted himself to analysing the existing data.

Brahe had set Kepler the task of analysing the pronounced departures of the planet Mars from the motion predicted using the Ptolemaic system.

As a believer in heliocentrism, Kepler's first thought was to match the motions to a mathematical scheme conceived by Copernicus but the results were unsatisfactory.

Knowing that the observations by Brahe were accurate, Kepler considered the situation afresh, accepting that the merit of any scheme he might devise would be determined by whether it matched the observations.

After many false starts, Kepler realised the data indicated that Mars travelled an elliptical path which had the Sun at one of the focal points of the ellipse instead of at the central point of a circle; the second focal point of the ellipse was vacant.

In fact, Copernicus had noted that Earth and the other planets revolve not around the Sun but about the *mean position* of the Sun. He had no idea why the Sun shouldn't be stationary.

Kepler realised that because the Sun was in a fixed position at one of the focal points of a planetary orbit, if the planet were considered to trace out a circular orbit (as Copernicus presumed from Aristotelian teaching) then the Sun would appear to oscillate about its mean position; that position being the centre of the circle. Hence if Copernicus had looked into the Sun's apparent movement more deeply he, rather than Kepler, might have discovered that planetary orbits are elliptical.

The speed of a body in a circular orbit would be uniform but it would vary for an ellipse. Kepler examined the variation in the rate of motion and established that Mars moved fastest when it was nearest the Sun and slowest when farthest away.

His task was to determine the mathematical relationship that would enable the speed of the planet to be calculated for any position on the ellipse. In the graphical presentation that Kepler used, the area within the ellipse that was swept out by the line that joined the planet to the Sun over any interval of time was directly proportional to that interval.

In 1609 Kepler published his analysis of the motions of Mars in a book entitled *Astronomia Nova* (*New Astronomy*).

Generalising from Mars, Kepler asserted that all of the planets, including Earth, were subject to two laws of planetary motion. The first law said that a planet describes an ellipse which has the Sun at one of its focal points. The second law expressed the rate at which a planet travelled, as described on the right.

Further analysis established the relationship between the scale of a planetary ellipse and the time taken to undertake a circuit of the Sun. In a second book *Harmonices Mundi* (*Harmony of the World*) published in 1619, Kepler announced his third law of planetary motion; that the squares of the times of revolution of any two planets around the Sun (including Earth) are proportional to the cubes of their mean distances from the Sun. He could determine this in the absence of knowledge of the true sizes of

RIGHT It was the pronounced 'looping' of Mars across the sky that enabled Kepler to devise his laws of planetary motion. (*Woods*)

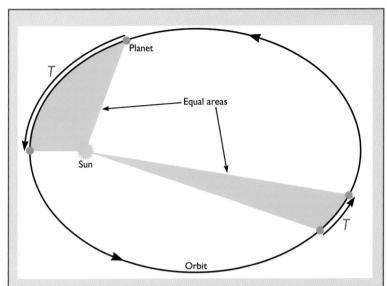

ABOVE The 'equal areas in equal time' law of planetary motion devised by Kepler. (*Woods*)

KEPLER'S LAWS OF PLANETARY MOTION

Johannes Kepler's first law of planetary motion states that the orbit of a planet is an ellipse that has the Sun at one of its two focal points.

The second law says the rate at which a planet travels varies, being fastest when nearest the Sun and slowest when farthest away. In the graphical presentation that Kepler employed, he was able to determine the speed of the planet at any position because the area in the ellipse that is swept by the line that joins the planet to the Sun over any interval of time is proportional to that interval. This is commonly described as the law of 'equal areas in equal times'.

The third law says that in relative terms, the squares of the times of revolution of two planets around the Sun are proportional to the cubes of their mean distances from the Sun.

These laws are empirical because although they state *how* planets orbit the Sun they don't explain *why* they do so.

Of course, these laws apply to Earth in orbit around the Sun, satellites orbiting planets, and the Moon orbiting Earth.

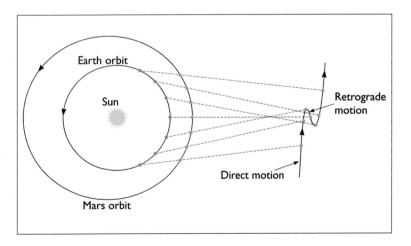

LUNAR LIBRATIONS

The rate at which the Moon turns on its axis is constant and its rotational period is synchronised with the time it takes to travel once around Earth. As a result, the Moon maintains a single hemisphere facing its primary.

Owing to the fact that the Moon's orbit is not circular but elliptical, its speed of progress varies. When the Moon is ahead of its mean position we gain a better view of the trailing limb, and when it is behind its mean position we gain a better view of the leading limb. This is libration in longitude.

A variety of smaller effects also influence the libration angles. In particular, perturbations from the Sun and the planets cause modulations. By taking all possible effects into account, the peak amplitude of the libration in longitude can vary in the range ±4° to about ±7° from the mean.

In addition, because the Moon's orbit is tilted relative to the Earth's equatorial plane, there is a libration in latitude. When the Moon is farthest south of the equator we gain a better view of its north pole, and when it is farthest north we gain a better view of its south pole.

The amplitude of the latitude libration is about ±6.5°.

Furthermore, as the Moon travels around its elliptical orbit, its distance from Earth varies and hence its apparent diameter varies. This results in a modest change in the size of the libration in latitude. When the Moon is nearest, we get a slightly better view of the northern and southern regions than when it is far away. Consequently the libration in latitude doesn't repeat on an exact cycle and it doesn't repeat with exactly the same amplitude from one month to the next.

Although at any given moment we are able to observe only a single hemisphere of the Moon, by taking into account all manner of libration effects it is possible, over time, to see 59% of the lunar surface.

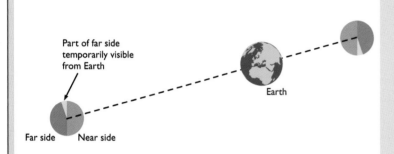

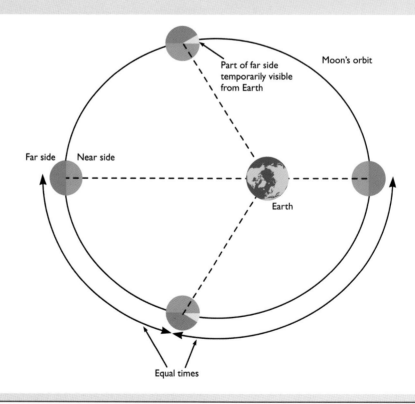

ABOVE AND RIGHT A depiction of the Moon's longitudinal librations caused by the manner in which its speed varies around its elliptical orbit, and latitudinal librations which derive from the fact that its orbit around Earth is inclined to the equator. *(Woods)*

the planetary orbits because the relationship depends only on their relative sizes.

By this mathematical analysis Kepler had fully explained the manner in which Mars traces out a 'loop' in the sky when opposite the Sun as viewed from Earth.

In his *Epitome Astronomiae Copernicanae* (*Epitome of the Copernican Astronomy*) which was published in three volumes between 1618 and 1621 Kepler provided an introduction to astronomy from the Copernican standpoint that explained both his own discoveries and those of Galileo.

Epitome generalised the conclusions which were previously published in relation to Mars to include all of the planets. Because the relative sizes of the long and short axes of their ellipses (known as the *eccentricity*) differed, the departures from idealised circular motion also differed. Knowing that the planets orbit the Sun in elliptical orbits also explained the fact – known since classical times – that the Sun does not travel across the sky on an annual basis at a constant rate. This arises because Earth itself travels around the Sun in an elliptical orbit.

As promotion of heliocentrism was prohibited by the Vatican, Kepler's *Epitome* was added to the list of banned volumes but it was well received in Protestant countries.

These laws also apply to the Moon, with Earth at one of the focal points of its orbit. Kepler realised that although the rate at which the Moon turns on its axis is fixed and is synchronised with its orbital period, the fact that it travels faster when at its closest point to Earth (which he named *perigee*) than at its farthest point (*apogee*) results in it sometimes leading and sometimes trailing the mean position in its orbit, enabling us to glimpse slivers of the far side at one or other equatorial *limb*. And similarly because the orbit of the Moon is inclined to the equator, when it is in the southern sky we can peer across its north pole and when it is in the northern sky we can observe its south polar region. These effects are known as *librations*.

Kepler coined the term *satellite* for an object that revolves around a planet, and of course this definition includes the Moon. After corresponding with Galileo about the Moon, Kepler gave us the terms *terrae* (land) and *maria* (seas) for the light and dark areas respectively.

To wrap up Kepler's story, after Rudolph II abdicated in 1611 and the new Emperor proved to have no interest in astronomy, Kepler went to Linz, Austria. A renewal of religious intolerance prompted him to flee back to Württemberg in 1626. He started to write a full account of the work of Brahe and himself but on his death from fever at the age of 58, little progress had been made.

Isaac Newton

One topic that had intrigued Kepler was *why* planets should travel around the Sun and moons around planets. He reasoned that there must be some direct coupling and for a while there was a suspicion that magnetism might be involved, but the actual cause was not realised until long after Kepler was dead.

Isaac Newton was born in Woolsthorpe, a village in Lincolnshire, on 25 December 1642. Unfortunately his father, who earned a modest living as a farm hand, had died three months previously. When his mother Hannah remarried, Newton was sent to live with his grandmother in nearby Grantham, where he began his schooling. Aged 16, he was called back to the farm to help his mother following the death of his stepfather, but he was of little assistance because his mind was distracted by intellectual pursuits.

In 1661 Newton enrolled at Trinity College in Cambridge. At that time in England, the works of Copernicus, Galileo, and Kepler had not been acknowledged by officialdom and it was widely believed that the Sun travelled around Earth on a daily basis. However, scientific societies had been established to discuss new developments, free of such constraints – most notably, in 1660, the Royal Society of London.

NEWTON'S DATE OF BIRTH

The 25 December 1642 date of birth for Isaac Newton is by the Julian calendar that England continued to use after Catholic countries adopted the Gregorian calendar in 1582 (according to which Newton was born on 4 January 1643). England didn't switch to the Gregorian calendar until 1752.

HOW ISAAC NEWTON USED THE MOON TO DETERMINE THAT GRAVITATIONAL ACCELERATION VARIES INVERSELY WITH THE SQUARE OF DISTANCE

The parameters that Newton had to work with in this study were the acceleration due to gravity at Earth's surface (g), Earth's radius (R_E), the distance to the Moon (D_M), and the time taken by the Moon to revolve around Earth (T_M).

We can demonstrate this using modern values:

$g = 9.8 m/sec^2$
$R_E = 6,375 km$
$D_M = 384,400 km$
$T_M = 27.3 day.$

Firstly, given the radius of the Moon's orbit, we can calculate its circumference as:

$2 \pi D_M$

where π is 3.14; this gives 2,415,256km.

Dividing the circumference by the time taken to make an orbit gives the speed of the Moon:

$2,415,256 / 27.3 = 88,471 km/day$

which equates to 1.03km/sec.

Knowing that a body that is constrained to travel at a constant speed in a circular path can do so only by accelerating towards the centre, Newton inferred that the radial acceleration of the Moon could be calculated as the square of this constant velocity divided by the distance to the centre.

This angular acceleration can be derived by the following logic.

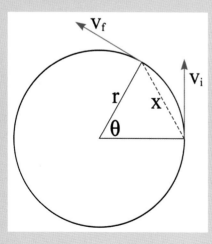

How Isaac Newton used the Moon to determine that gravitational acceleration varies inversely with the square of distance

For a small angle (θ) the distance (x) around the circle is:

$x = r . \theta$

and hence

$\theta = x / r .$

When travelling at velocity v for a given interval (Δt), we can say:

$x = v . \Delta t$

and by substituting:

$\theta = v . \Delta t / r .$

The difference in velocity (Δv) is the final velocity minus the initial velocity:

$\Delta v = v_f - v_i$

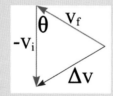

which is directed towards the centre of the circle. For a small θ in the velocity triangle:

$\Delta v = v . \theta$

and hence:

$\theta = \Delta v / v .$

Equating for θ:

$\theta = \Delta v / v = v . \Delta t / r .$

and by rearranging:

$\Delta v / \Delta t = v^2 / r$

which is the radial acceleration (a_r).
So plugging in the numbers:

$$(1.03 \times 1.03) / 384{,}400 = 0.0000027$$

measured in km/sec^2. For our purposes this is better expressed as 0.0027m/sec^2 for direct comparison with the acceleration due to gravity at Earth's surface.

If the gravitational force exerted by Earth can be treated as if it originated at the centre of the planet then a measurement of the acceleration due to gravity at the surface (g) is 6,375km from that point and the Moon lies 60 times farther away.

The task Newton had set himself was to work out how the acceleration imposed by gravity diminished with distance. He now had two values for that force: one at Earth's surface and the other at 60 times that distance. The radial acceleration clearly wasn't inversely proportional to distance because in that case the acceleration at lunar distance would be 1/60th that at Earth's surface. In fact, 9.8 = 0.0027 × 60 × 60

means it is inversely proportional to the *square* of distance. Of course, we're using modern values for these quantities and they weren't all known with such precision in Newton's time.

Galileo was first to measure the acceleration due to gravity at Earth's surface. This value had been refined by his successors. The time taken by the Moon to revolve around Earth was well-known. Hipparchus had correctly estimated the distance to the Moon in terms of the radius of its primary. But the size of Earth's radius used by Newton in his early calculation was 15% out. Even so, the result strongly hinted that gravity is inversely proportional to the square of an object's distance from Earth's centre.

Eager students were stimulated by this 'buzz'. On graduating in 1665 Newton had absorbed all the mathematics that his tutors had to offer and was developing his own ideas.

Bubonic plague struck London in 1665 and claimed over 30,000 people during the summer. In response to this spreading epidemic, Trinity College closed in August and Newton, who had just started on his master's degree, retreated to Woolsthorpe and spent the next two years working on his own ideas.

Newton made tremendous advances during this time, but once he had figured something out to his own satisfaction he put it aside and never told anyone. For example, on devising the mathematics that he named the method of fluxions and we know as calculus, he kept it to himself. He had a long and productive career, but here we need concern ourselves only with gravity.

In 1666, while in isolation, Newton began to wonder about celestial bodies. He was aware that Galileo and Kepler had explained *how* they moved but he wanted to know *why* that was the case. After several weeks of musing, one moonlit night he was sitting beneath a tree in the farm's orchard and heard the soft thud of a falling apple striking the ground nearby. (Contrary to myth, the apple didn't strike his head.) Once the apple had become separated from its twig, something had caused it to fall to the ground. Why didn't the Moon fall?

The *existence* of gravity as the force which caused objects to fall to Earth was well known, but its *nature* was mysterious, in that whereas every other instance of applying a force to cause an object to move required there to be physical contact, gravity was different because it acted at a distance.

Newton pondered Galileo's reasoning that an idealised object, once in motion, would travel in a straight line unless a force intervened to change its course. What was it that prevented the Moon from travelling in a straight line?

Conventional wisdom said that for the Moon to travel in a curved path around Earth it must be subjected to some force that continuously acted *in the direction of travel*. Newton's epiphany was to conjecture that the force originated from Earth and acted *perpendicular* to the direction of travel at all times.

If this force was gravity then its influence

must extend to the celestial realm, and it seemed sensible that its strength would diminish with increasing altitude. Furthermore, because Earth is spherical its gravitational attraction can be deemed to originate from a point at its centre. Newton therefore measured the strength of the force at Earth's surface, for which he needed to know Earth's radius, and then investigated different possible relationships for how much weaker it would be at the Moon's distance. He decided the strength of the force was inversely proportional not to the separation between the two objects but to the *square* of that distance. Because the radius of Earth was not precisely known, his result wasn't conclusive and he put the calculation aside.

In 1669 Newton was awarded the Lucasian chair of mathematics at Cambridge and three years later was elected a Fellow of the Royal Society.

On hearing in 1679 of a revised estimate of the radius of Earth, Newton repeated his earlier calculation and found a very good match with an inverse square relationship.

The next task was to determine what the path of the Moon would be under the influence of such a force. The natural tendency of the Moon was to travel in a straight line, one that was tangential to the line which connects its centre to Earth's centre. But the attractive force along that radial line causes the Moon to accelerate Earthward. Further calculation established that the Moon travels around Earth in an ellipse. Newton then used Kepler's laws to demonstrate that the same reasoning applied to planets orbiting the Sun, showing gravitation to be a universal force.

A crucial step was to introduce the notion of the intrinsic *mass* of an object; as distinct from its *weight*, which was a measure of the force exerted on it by gravity. Although the Moon has tremendous mass, it is *weightless* by being in a state of perpetual *free fall* around Earth. By generalising, Newton stated that the gravitational force between two bodies is directly proportional to the product of their masses and inversely proportional to the square of the distance separating their centres. To provide the scale there was a factor which he called the *gravitational constant*. This had to be determined by experiment and proved difficult

ABOVE A portrait of Isaac Newton painted in 1689 by Godfrey Kneller when Newton was 46 years of age and shortly after his *Principia* was published.

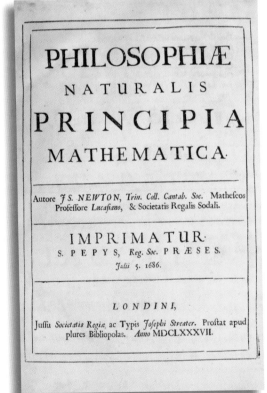

RIGHT The title page of Isaac Newton's *Principia*.

EDMVND. HALLEIVS LL.D.
GEOM. PROF. SAVIL. & R.S. SECRET.

to measure; the first successful measurement was obtained by Henry Cavendish in 1798 when he, in effect, weighed Earth.

Despite having discovered the truth about celestial motions, Newton didn't tell anyone; he simply put his calculations aside and turned to another line of research.

But others had begun to give thought to the nature of gravity, and in 1684 Newton received a visit from astronomer Edmund Halley who, with colleagues at the Royal Society, *suspected* it to be a force with an inverse square relationship. What they wanted to know, but were unable to calculate, was the motion of the Moon under a force of this kind – and in particular whether it would be an ellipse as proposed by Kepler. To Halley's astonishment, Newton said that he had already proved that it would be elliptical.

On realising the scope of Newton's research into gravitation, Halley urged him to publish his results. Newton said he would but the task took several years, largely because he had to start by explaining the mathematical tools which he had invented. The result, published in 1687, was a tome called *Philosophiæ Naturalis Principia Mathematica* (*Mathematical Principles of Natural Philosophy*).

The first law of motion was the principle of inertia. Based on Galileo's insight, this said that an object will continue in its state of rest or of uniform motion in a straight line, except in so far as it is compelled by an applied force to alter its state. The second law said that the force acting on an object is equal to the product of its mass and the resultant acceleration. The third law said that for every action on an object there must be an equal and opposite reaction.

Newton was gracious in acknowledging his predecessors, saying that he had 'stood on the shoulders of giants' such as Galileo and Kepler. By accounting for almost all celestial motions, Newtonian mechanics has enabled us to send spacecraft to investigate the planets.

NEWTON'S LAWS OF MOTION

Newton's three laws of motion founded classical mechanics by describing the relationship between a body and the forces acting upon it, and its motion in response to the applied forces.

The first law states that an object will either remain at rest or continue to move at a constant velocity unless acted upon by an external force.

The second law states that the vector sum of the external forces acting on an object is equal to its mass multiplied by its acceleration vector.

The third law states that if one body exerts a force on another body, that second body acts to exert an equal and opposite force upon the first body.

In his *Principia*, Newton demonstrated that his three laws of motion, combined with the law of universal gravitation, fully explained Kepler's empirical laws of planetary motion.

LIMITATIONS OF NEWTON'S LAWS

The limitations of Newton's laws of motion only became evident when Albert Einstein developed the concept of relativity.

LICK OBSERVATORY, MT. HAMILTON, CAL.

Chapter Four

The Moon in the sky

This chapter explains how and why pre-Space Age telescopic observers mapped the Moon, and the role that the 'longitude problem' played in this effort.

OPPOSITE The Lick Observatory constructed in the 1880s atop Mount Hamilton, near San Jose in California, was the world's first high-altitude astronomical observatory. This picture was taken circa 1900. *(Lick Observatory)*

In the 2nd century BC, Hipparchus introduced a system of latitude and longitude coordinates to specify geographical positions.

Determining a latitude was a straightforward task because it was based on the elevation of the Sun at local noon. Longitude was much more difficult. After the longitudes of several sites had been determined by measuring the times of eclipses and other astronomical phenomena, it was possible to compile a map by estimating the positions of other locations relative to these known points. It was straightforward to navigate at sea provided that a ship remained close to a known coastline, but when out of sight of land navigation was perilous.

The 'longitude problem'

The 'longitude problem' was a vital issue for Christopher Columbus, who crossed the Atlantic from Spain in 1492 in search of a sea route to the Indies and instead discovered various islands in what would be named the Caribbean Sea. Also for Vasco da Gama who sailed around Africa in 1498 to finally establish a sea route to India. Within a generation, ships had circumnavigated the world.

Early ocean navigators used the process of

RIGHT Johannes Müller, known as Regiomontanus, portrayed by Bräht. *(Smithsonian Institution).*

'dead reckoning' by estimating the heading and speed of their ship over a given interval, but this became progressively less accurate. In order to reduce the risk of getting lost, navigators would often sail to the latitude of their destination, then turn in the desired direction and follow a line of constant latitude. Although safer, this denied a ship the route which offered the most favourable winds and currents. What was required was a method of accurately determining longitude on the open ocean to facilitate optimum routes.

Since Earth rotates on its axis at the steady rate of 360°/day, there is a direct relationship between time and longitude. If a navigator knew the time at a fixed reference point when making an observation at sea, the difference between the reference time and the apparent local time would enable the ship's position to be calculated relative to that fixed location. Finding apparent local time was relatively easy. The problem was to know the time at a distant reference point, for example Greenwich in England, at that same moment.

Shortly prior to his death in 1474, aged 40, Johannes Müller of Königsberg, who had translated Ptolemy's *Amalgest* from Greek to Latin, realised that if the future path of the Moon could be determined with sufficient accuracy to enable its position to be predicted for a number of locations on land then observations of the position of the Moon relative to the stars by a ship at sea should enable the ship's position to be calculated. Although this process was feasible in principle, it wasn't practicable at that time.

Nevertheless, over the years other schemes were devised to exploit observations of the Moon in solving the 'longitude problem'.

Early mapping of the Moon

Jacob Floris van Langren founded a business in Amsterdam in 1586 that made globes, and as Dutch explorers made discoveries he was barely able to keep up with the demand for updates. In 1627 his grandson, Michel, observed the Moon and made a sketch. After moving to Madrid as Court Astronomer to King Félipe IV of Spain in 1630, Michel said that tables listing the sunrise and sunset times of given lunar features would provide terrestrial observers with the means to calculate the actual time for use in determining their longitude.

The first step towards achieving this was to make a map of the Moon. In 1643, having made 30 sketches, van Langren realised that he had competitors, so in 1645 he issued a whole-disc map 34cm in diameter with 325 features named after prominent philosophers, mathematicians, astronomers, explorers, religious figures, and (in honour of his sponsor) Spanish royalty. But the revolt of Protestantism that would later be called the Thirty Years War was underway and a nomenclature which was drawn from Catholic Europe was sure to be contentious. Interestingly, one of the names that

survived was the prominent crater Langrenus, the Latinised form of his surname, by which he honoured his family.

In 1637 Pierre Gassendi, a mathematician in Paris, also reasoned that it ought to be possible to use the Moon to determine the time. After he had made several drawings, he heard that Johannes Hevelius, whom he had once met, was making a map, and upon seeing the high quality of the younger man's sketches Gassendi handed over his own work.

A city councillor in Danzig in the Polish-Lithuanian Commonwealth, Hevelius built an observatory onto his residence and installed a telescope with a 5cm lens, a focal length of

RIGHT The map of the
Moon published by
Hevelius in his book.

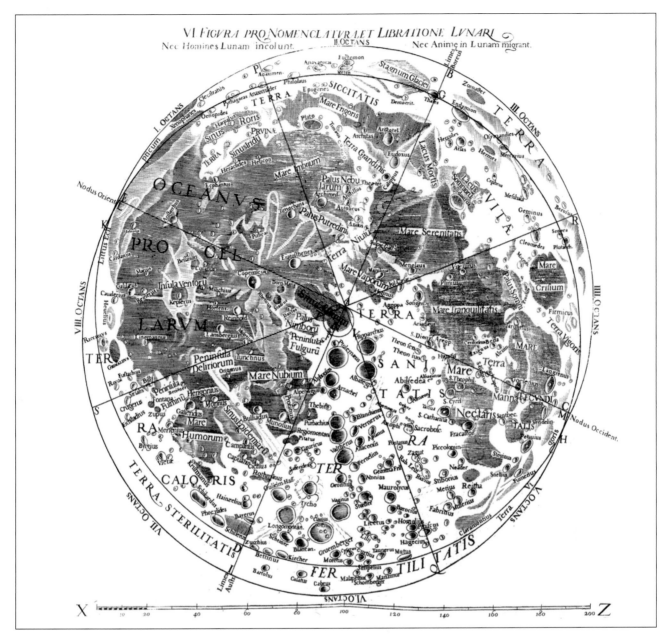

3.6m and a magnification factor of 50; it was one of the best telescopes of the time.

Selene was the Greek goddess of the Moon, so Hevelius called his 1647 book about the lunar features *Selenographica sive Lunae Descriptio* (*Selenography, or A Description of The Moon*). It was illustrated with his own sketches and included a consolidated map 30cm in diameter. This was the first to depict the bright streaks of the lunar 'rays' that radiate out from some craters. Like Galileo, Hevelius estimated the heights of the lunar peaks by their shadows but with more accurate results.

As a Protestant, Hevelius's nomenclature had little in common with that of van Langren. In fact, he had presumed himself to be the first to name features and described the task as arduous. His 275 names were inspired by terrestrial landforms, including oceans, seas, bays, and lakes, but only a few of his names have survived.

Giovanni Battista Riccioli, a Jesuit professor of astronomy and theology at the University of Bologna, was a firm believer in the Aristotelian system of the heavens as described by Ptolemy. To counter the increasing popularity of Copernican heliocentrism he set out to write an authoritative history of astronomy. However, in developing his argument Riccioli came

ABOVE Riccioli's map of the Moon, published in *Almagestum Novum*.

ABOVE Cassini's map of the Moon.

of Hevelius, its historical significance was the nomenclature. It retained oceans, seas, and bays for the dark areas but renamed them for states of mind; for example the Ocean of Storms, the Sea of Tranquillity, and the Sea of Serenity. The craters were assigned the names of astronomers and philosophers associated with the study of the Moon, including both Riccioli and Grimaldi. The despised Copernicus was assigned a crater which, although prominent, was in the Ocean of Storms. Galileo the heretic was assigned only an obscure crater. To Hevelius's frustration, soon copies of *his* map were in circulation relabelled in this manner! Almost all of the 200 names of Riccioli's scheme remain in use today.

Giovanni Domenico Cassini was born in 1625 in the Republic of Genoa. After gaining a Jesuit education he was hired by the Marquis Cornelio Malvasia in Bologna and made observations to support his employer's interest in astrology. In 1650 he was appointed professor of astronomy at the local university.

When the French government decided that a national observatory should be created in Paris, Cassini, who had made a number of significant astronomical discoveries, was invited by Louis XIV to serve as its director. Cassini relocated to Paris in 1669 to oversee the construction of the observatory, which upon its completion in 1671 possessed a variety of instruments.

In 1679 Cassini published a map of the

to suspect that geocentrism was wrong! Of course, he couldn't admit this publicly.

Riccioli's treatise *Almagestum Novum* (*New Almagest*) was published in 1651 and featured a whole-disc map of the Moon 28cm across which was based on observations made by his student, Francesco Grimaldi.

Although this map was little better than that

RIGHT Giovanni Domenico Cassini at the Paris Observatory. *(Bibliothéque nationale de France, Paris)*

FAR RIGHT A portrait of John Flamsteed by George Vertue.

Moon which, at 52cm wide, was rather larger than those of his predecessors. It was very accurate, but so few copies were released that his map didn't gain the attention it warranted.

Although interest in the surface of the Moon waned for a time, interest in its celestial motion picked up.

Solving the 'longitude problem'

In 1674 King Charles II of England appointed a Royal Commission to investigate a scheme for solving the 'longitude problem' which was similar to that proposed by Johannes Müller. When this Commission reported, it noted that the first task was to compile an accurate map of the star fields through which the Moon passed – a substantial portion of the sky, owing to the inclination of the Moon's orbit to Earth's equator. In 1675 the Royal Observatory at Greenwich was established for this purpose and John Flamsteed, having assisted the Commission, was appointed as the first Astronomer Royal in order to oversee the drawing up of a catalogue of 3,000 stars.

The other aspect of the task was to precisely define the motion of the Moon. Isaac Newton

BELOW A portrait of Pierre-Simon Laplace by Jean-Claude Naigeon.

ANOMALOUS MOTIONS OF THE MOON

Evection is the largest anomaly produced by the Sun in the monthly orbit of the Moon around Earth. It was discovered by Ptolemy during a detailed examination of the work of Hipparchus.

In studying the motion of the Moon, Hipparchus had concentrated on historical observations of eclipses, which occur only at the 'new' and 'full' phases. Ptolemy made his discovery while checking his predecessor's conclusions against observations of the Moon taken at other times in phase cycle. He found a considerable discrepancy at either 'first quarter' or 'last quarter'. To accommodate this Ptolemy ingeniously added another circle to his scheme of an epicycle and a deferent. Although this complicated the scheme, it had the virtue of better accounting for the motion of the Moon.

We now know this effect to be a result of Earth's orbit being elliptical. That is, the varying strength of the Sun's gravitational pull imposes a cyclical variation in the eccentricity of the Moon's orbit, with a resultant variation in the position of its perigee.

Tycho Brahe discovered an anomaly which vanished at 'new', 'full', 'first quarter', and 'last quarter', but was significant at points in between. He rather mundanely named this *variation*. It is a speeding-up of the Moon as it approaches 'new' and 'full', and a slowing-down as it approaches 'first quarter' and 'last quarter'.

Later on, Brahe found a small anomaly which depended upon the time of year, so he named it *annual inequality*.

Isaac Newton qualitatively explained this in terms of the orbit of the Moon becoming slightly expanded in size and lengthening in period when Earth is at its closest to the Sun in early January, when the influence of the Sun is strongest; and slightly reducing in size and shortening in period when Earth is at its farthest from the Sun in early July, when the influence of the Sun is at its weakest.

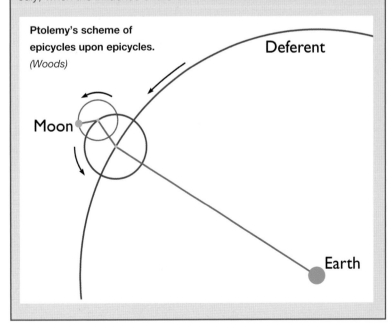

Ptolemy's scheme of epicycles upon epicycles.
(Woods)

had used his insight into gravitation to explain why the Moon revolves around Earth, but his knowledge of its motion was rudimentary. Even after allowing for *evection* (discovered by Ptolemy) and for *variation* and *annual inequality* (both discovered by Tycho Brahe) the Moon didn't precisely follow his predicted path. In particular, he was unable to accurately predict the time of perigee.

In analysing records of eclipses dating back to Hipparchus in the 2nd century BC, Edmund Halley found in 1695 that the Moon was slowly speeding up, a property that became known as *secular acceleration*. In 1749 this anomaly was calculated by Richard Dunthorne at Cambridge University at 10 seconds of arc per century. Its cause was not known.

Pierre-Simon Laplace was born in 1749, the son of a farm labourer in Normandy. Noting his intelligence, wealthy neighbours ensured that he received an education. He went on to become one of France's leading mathematicians.

Laplace announced in 1786 that his analysis had revealed that the gravitational influences of the other planets in the solar system were causing the eccentricity of Earth's orbit around the Sun to decrease (making it less elliptical, more circular). This partly explained the secular acceleration of the Moon, but the remainder was still a mystery.

To jump ahead on this issue, in the late 19th century George Darwin, a son of the renowned naturalist, realised that the transfer of angular momentum from Earth to the Moon was both slowing the rate at which the planet spins (thus lengthening the day) and accelerating the Moon in its orbit (causing it to recede and lengthening the month).

As more influences were taken into account to improve the predicted motion of the Moon it was realised that there was an entirely different problem: the rotation of Earth on its axis wasn't uniform. The planet isn't perfectly spherical with a uniform distribution of mass. In addition, there is geological activity. To conserve angular momentum over the short term, its rotation rate varies in an irregular manner. At the finest levels of accuracy therefore, the rate at which Earth spins is sufficiently variable to cause the predicted position of the Moon to differ from its true position. This doomed efforts to use the Moon as an accurate clock. However, it was no longer necessary.

In 1714 the British government had decided to establish a Board of Longitude to administer prizes that were intended to encourage innovators to solve the problem of finding longitude at sea. The result was the invention in 1761 by John Harrison of an accurate chronometer suitable for use by mariners.

The Moon's orbit

The path of the Moon in the sky is inclined to the ecliptic. Hipparchus was the first to measure this angle as about 5°. The nodes where the two planes intersect drift in a retrograde manner (i.e. from east to west) around the ecliptic, completing a circuit of the sky in 18.6 years.

Tycho Brahe found that the inclination of the

BELOW The rotation of the Moon's line of apsides. *(Woods)*

BOTTOM The synodic month is longer than a sidereal month because the Earth–Moon system is travelling around the Sun. *(Woods)*

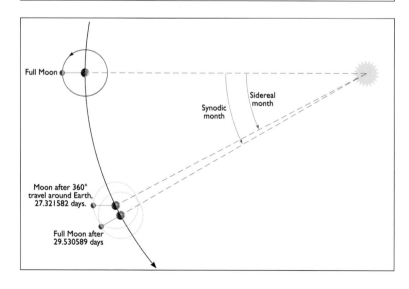

Moon's orbit with respect to the ecliptic isn't fixed; in fact, it oscillates in a regular manner. Consequently the motion of the nodes is also variable. The angle is currently 5.145°, but it can vary between 4.98° and 5.30°.

Given that the Moon's orbit is inclined to the ecliptic and the ecliptic is inclined to the equator, the tilt of the Moon's path relative to the equator changes over the course of an 18.6- year cycle. At its maximum, the Moon passes 28.6° north and south of the equator as it pursues its orbit of Earth, and at its minimum tilt the Moon remains within 18.3° of the equator.

The motion of the Moon in the sky doesn't precisely match what would happen if its orbit were a circle, because sometimes it gets ahead of schedule and at other times it slips behind.

Hipparchus was the first to observe that the part of the Moon's path in which its motion was the fastest wasn't always in the same position on the sky.

The positions of fastest and slowest speed migrate eastward along the Moon's path. These points are necessarily 180° apart and the line through the focal points that links them is called the *line of apsides*. The apsides take 8.85 years to complete a circuit, travelling from west to east.

It was Kepler who realised the rotation of the line of apsides was the result of the Moon's orbit being an ellipse.

Perigee is currently 362,600km but it varies between 356,400km and 370,400km. Apogee is 405,400km and varies between 404,000km and 406,700km. The mean distance (half of the line from perigee to apogee, called the *semi-major axis*) is currently 384,400km.

The month is a tricky concept and its length depends on what is being measured.

The Moon completes an orbit around Earth in 27.321582 days with respect to the stars, and this is the *sidereal* month.

But because the Earth–Moon system is travelling around the Sun, every month the Moon has to travel a little more than 360° in order to catch up with the progress of the Sun in the sky and establish the same illumination phase. This lunar cycle is 29.530589 days and is called the *synodic* month.

The interval from one perigee to the next is called the *anomalistic* month. This arises because the line of apsides is rotating around

the orbit of the Moon and therefore the Moon must travel a little further in order to catch up. Although it has a wide variation, its mean length is 27.554551 days.

The shape of an ellipse is defined by the parameter called *eccentricity*, with a higher number indicating a more extended shape. The eccentricity of a circle is zero.

BELOW Mayer's map of the Moon.

T. Mayer del.   J.P. Kaltenhofer sculp. Gottingae.

RIGHT Johann
Hieronymus Schroeter.

FAR RIGHT The title
page of Schroeter's
Fragmente.

Earth's orbit around the Sun has an
eccentricity of 0.0167, hence it is only slightly
non-circular.

The variation in the gravitational attraction
of the Sun while the Earth–Moon system
completes its annual circuit perturbs the Moon's
orbit, causing its eccentricity to vary between
0.023 and 0.078 with an average of 0.055.
As a result the orbit of the Moon doesn't
precisely repeat and the actual interval between
consecutive perigee passages (the anomalistic
month) can be as short as 24.7 days or as long
as 28.5 days.

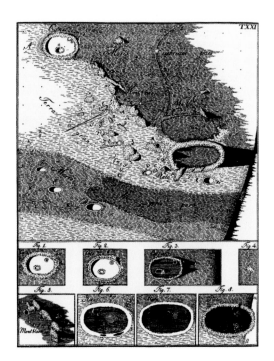

RIGHT A page from
Schroeter's book
featuring the lunar
Alps and the dark-
floored crater Plato.

A renewal of lunar mapping

As regards the Moon, an interest had been
rekindled in studying its surface.

When Tobias Mayer in Nuremberg decided
in 1748 to map the Moon, he made his first
task to precisely determine the line of the
lunar equator. After two years he published his
observations of the limb effects of libration.

As a skilled draughtsman, Mayer used
an eyepiece micrometer to measure the
geographical positions of lunar features in
order to make a full-disc map, but this wasn't
published until 1775; more than a decade
after his death. Although only 20cm wide, it
was the first map to have lines of latitude and
longitude. It superseded the map by Hevelius
(re-annotated with Riccioli's nomenclature) that
had been definitive for 125 years and it wouldn't
itself be surpassed for half a century.

Johann Schroeter was born in 1745 in Erfurt
in Germany, and upon qualifying in law at the
University of Göttingen in 1767 he moved to
Lilienthal near Bremen in 1781 to become
the chief magistrate. Having been inspired by
Mayer's map, he built an observatory alongside
his house and installed a series of ever-
improved telescopes. Although an accurate
observer he wasn't a draughtsman, so he
employed a 'schematic' style.

Over a period of 30 years Schroeter produced
hundreds of drawings of individual lunar features
at high magnification under different angles of

FAR LEFT Johann
Heinrich von Mädler.

LEFT Wilhelm Wolff
Beer. *(Wikipedia)*

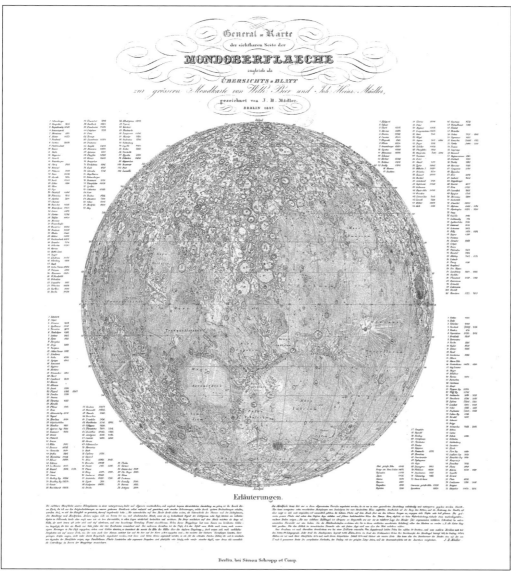

LEFT The whole-disc
map of the Moon
published by Beer and
Mädler in their book
Der Mond.

illumination. In this respect he founded modern selenography. He was the first to study the canyons (called *rilles*) that come in linear and sinuous varieties. His measurements of the heights of lunar mountains were more numerous and far better than those of his predecessors.

In 1791 Schroeter published his masterpiece *Selenotopographische Fragmente zur genauern Kenntniss der Mondfläche* (*Fragments of Lunar Topography illustrating the Nature of its Surface*). These two volumes were illustrated by exquisite engravings of shaded-relief drawings. Unlike his predecessors, Schroeter didn't integrate his work into a full-disc map. Instead, he used an enlarged form of Mayer's map to provide context for his readers. He also extended Riccioli's nomenclature by introducing a scheme for naming the subsidiary features that were in the vicinity of already named features.

Schroeter inferred that if the Moon possessed an atmosphere, then its pressure was less than that of the best vacuum pump available at that time. One of his long-term objectives was to discover whether the Moon's surface underwent changes; he found no evidence to suggest this.

Unfortunately, in 1813 Napoleonic soldiers ransacked Schroeter's home and most of his unpublished work was lost.

Johann von Mädler was born in 1794 and became a teacher in Berlin. One of his students, the wealthy banker Wilhelm Beer, was only a few years younger and they became friends. In 1829 Beer constructed an observatory at his home, bought a state-of-the-art telescope that had a 100mm lens and a micrometer, and hired Mädler as observer. They promptly set out to create what they intended to be the definitive lunar map.

The first task was to define a reference grid by using surveying techniques in order to measure surface features relative to a set of control points. After 600 nights at the telescope they had almost 1,000 objects trigonometrically defined. Starting in 1834 they published *Mappa Selenographica* in four parts, one part per annum. Together these formed a whole-disc map which was 95cm across. They borrowed names from the nomenclatures of Hevelius, Riccioli, and Schroeter, and added 140 more names plus their own scheme for naming subsidiary features; a scheme that would later provide the basis for the one we use today.

In 1838 Beer and Mädler reissued their map with a dissertation in the book *Der Mond* (*The Moon*).

So thoroughly had Beer and Mädler done their

RIGHT John William Draper.

FAR RIGHT William Cranch Bond. *(Harvard College Observatory)*

job that their map remained the definitive work until late in the 19th century, during which time people concentrated on investigating the fine detail which was below the resolution of the map.

Early photography of the Moon

The next advance in the study of the Moon came with the advent of photography.

The first picture of the Moon was taken in 1840 by J. W. Draper in America. He arranged a small lens to focus a lunar image onto a light-sensitive plate and then employed a clockwork mechanism to compensate for the substantial drift of the Moon across the sky during the 20min exposure. The result was crude, but it showed recognisable features.

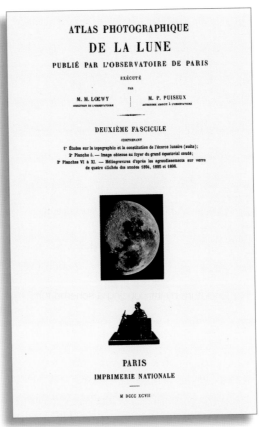

RIGHT William Henry Pickering. *(Harvard College Observatory)*

FAR RIGHT Francis Gladheim Pease. *(Mount Wilson Observatory)*

BELOW Pickering's book **showed areas of the moon under five illumination phases.**

W. C. Bond in America was a self-educated watchmaker who developed an enthusiasm for astronomy after he witnessed a solar eclipse in 1806, aged 19. He created a private observatory which was the best in the country. In view of his work, he was invited in 1847 to become the first director of the Harvard College Observatory.

In pioneering astrophotography, in December 1849 Bond used a 38cm refracting telescope to obtain a plate of the Moon in 20min that showed all the principal features.

The invention of a much faster technique for capturing an image enabled Warren de la Rue in England to take a telescopic image of the Moon in 1852 with an exposure of less than a minute. Although the lunar disc was only 28mm wide, it was sufficiently sharp to enable it to be enlarged to match the size of some of the early whole-disc sketch maps.

The introduction of *dry emulsion* in the late 1870s facilitated detailed photographic studies of the Moon.

At the Paris Observatory, Maurice Loewy and his assistant Pierre Puiseux utilised the innovative coudé refractor to take photographs

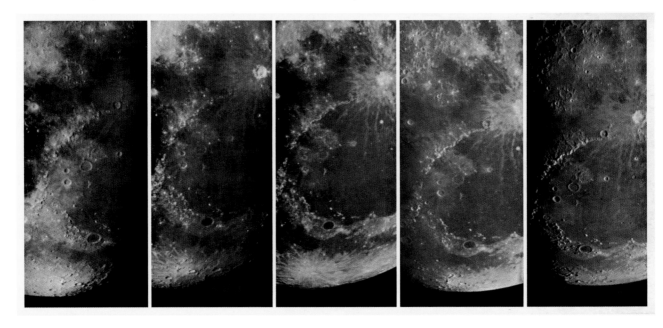

for *Atlas Photographique de la Lune*. It was published in 1896 and covered the disc in small sections that were scaled such that if a full disc had been created it would have spanned about 85in.

In 1897 the *Lick Observatory Atlas of the Moon* by E. S. Holden was published in America using pictures taken by the 92cm refracting telescope. In this case the Moon was presented on the same scale as Beer and Mädler's map.

After erecting a temporary astronomical facility on Jamaica for the Harvard College Observatory, W. H. Pickering set out to photograph the lunar disc in sections for each of five illumination phases. Although his images weren't as good as those of his competitors, Pickering featured them in his 1903 volume *The Moon: A Summary of Existing Knowledge of our Satellite, with a Complete Photographic Atlas*.

In commissioning a telescope at the Mount Wilson Observatory in California, whose 2.5m mirror was the largest in the world at that time, Francis Pease decided to perform a test by photographing the Moon during three nights in September 1919, which he accomplished with spectacular results.

In general though, turbulence in Earth's atmosphere, which astronomers refer to as *seeing*, made pictures of the Moon sufficiently blurry that there remained scope for visual studies, particularly in the limb regions.

When 20th-century professional astronomers turned their attention to the stars and even more distant objects, they came to regard the Moon as a source of 'light pollution' and left it to their amateur brethren.

The International Astronomical Union was established in 1919 to oversee general issues, and it took responsibility for regulating lunar nomenclature. It created a committee to rule on conflicting claims and in 1935 published a list of features on a map that followed Beer and Mädler's scheme.

LEFT The Hooker 2.5m reflecting telescope of the Mount Wilson Observatory. *(Mount Wilson Observatory)*

RIGHT This picture taken in 1919 by Francis Pease includes the Sea of Clouds and the craters Ptolemaeus and Alphonsus at the bottom and Tycho at the top. *(Mount Wilson Observatory)*

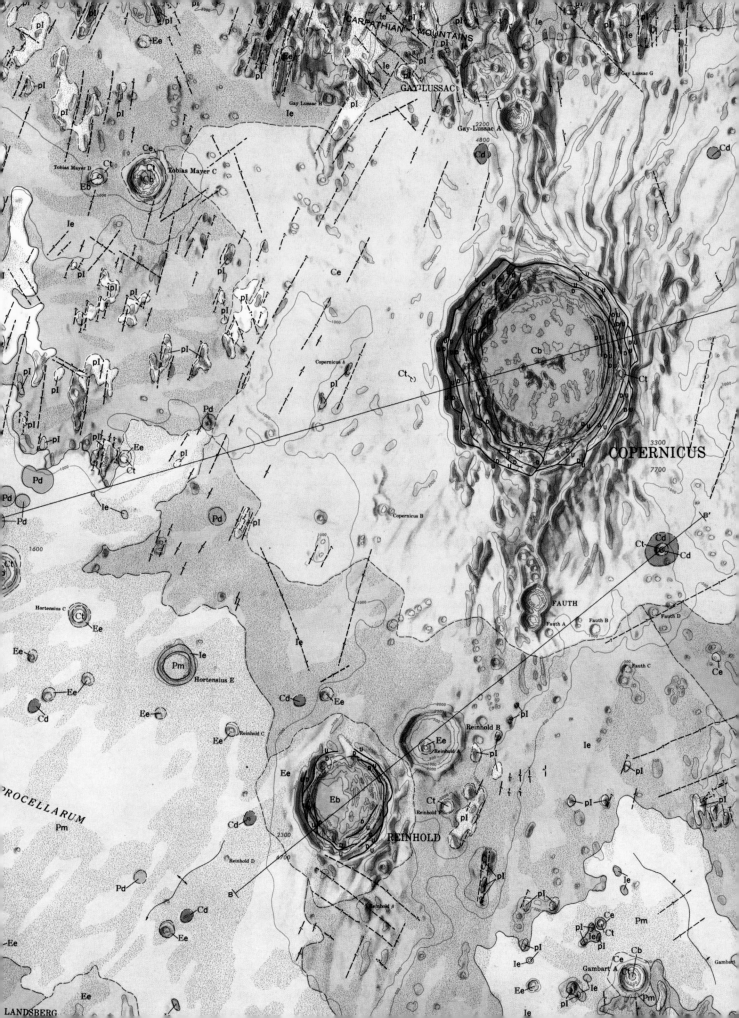

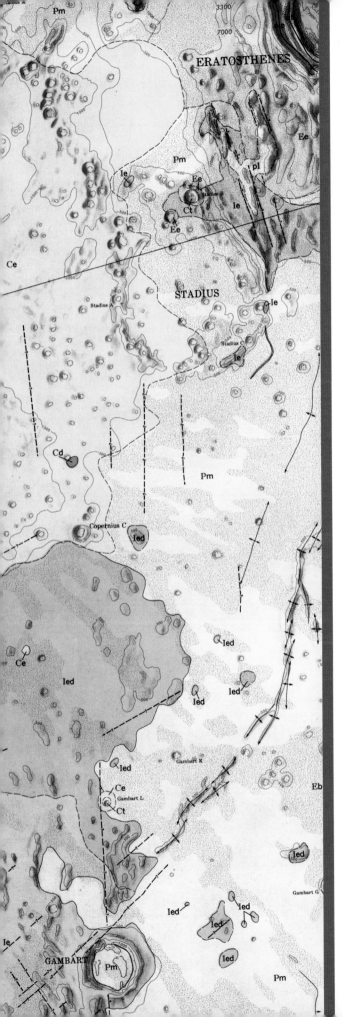

Chapter Five

The origin of
lunar craters

This chapter reviews pre-Space Age investigations of the nature of the lunar surface and the processes suspected of being involved in the creation of its features.

OPPOSITE The prototype geological map made by Gene Shoemaker in 1960 by inferring the superposition relationships around the crater Copernicus using a picture taken in 1919 at the Mount Wilson Observatory. *(USGS Astrogeology Science Center)*

In 1662 Robert Hooke was made curator of the recently formed Royal Society of London and charged with devising demonstration experiments. As an extremely skilled technical artist, he published a book in 1665 that included a detailed sketch of the lunar crater Hipparchus, located at the centre of the Moon's disc. He had no wish to map the Moon, but instead carried out experiments to investigate how the craters might have formed. First he dropped heavy balls into wet clay. He also heated alabaster until it bubbled, then let it set so that the final bubbles to break the surface made craters. However, just as Hooke couldn't conceive of where the projectiles could have come from to scar the Moon so intensively, neither could he imagine how the surface could ever have become hot enough to blister on such a scale.

Nevertheless, the prevailing view was that the craters must have been produced by volcanism in some manner.

Experimenters

James Carpenter, an astronomer at the Royal Observatory at Greenwich in England, and James Nasmyth, an engineer, carried out experiments to study how lunar features may have been formed. They concluded that the craters were produced by volcanic 'fountains'. In the early phase of such an eruption, when the velocity of the material ejected from the vent was great, the material would spray out in an umbrella-like plume and fall back some distance away to build up a concentric ring which became the wall of the crater. In some cases, as the eruption declined, the fallout would create a succession of terraces interior to the wall. The final phase of the eruption might construct a central peak.

For craters having dark floors and no peak, it was presumed the eruption switched to fluid lava that buried the vent. Given the weak lunar gravity and the lack of atmosphere, it seemed plausible that this process could have produced very large structures.

In addition, Nasmyth made a 50cm reflecting telescope with which he and Carpenter studied a number of lunar features in detail. They used the observations to produce exquisite models

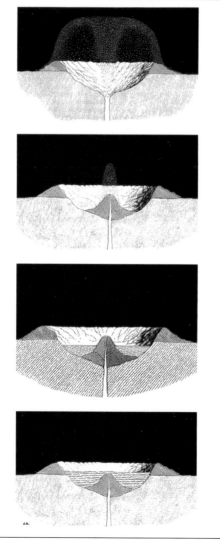

RIGHT The 'fountain'
concept for the
formation of lunar
craters proposed
by Nasmyth and
Carpenter. After the
eruption accumulates
a peripheral ring
structure (top), the
pressure in the vent
declines and the
remaining material
forms the central peak
(upper middle). Lava
can then partially
fill the cavity (lower
middle) and in some
cases bury the vent
(bottom). Adapted
from *The Moon:
Considered as a
Planet, a World, and a
Satellite*.

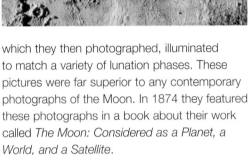

which they then photographed, illuminated to match a variety of lunation phases. These pictures were far superior to any contemporary photographs of the Moon. In 1874 they featured these photographs in a book about their work called *The Moon: Considered as a Planet, a World, and a Satellite.*

A new perspective

G. K. Gilbert was born in Rochester, New York, in 1843. After making a number of surveys as a field geologist, he was appointed Senior Geologist when the US Geological Survey was established in 1879.

Over a period of 18 nights in August through to October in 1892 Gilbert examined the Moon using the 65cm refractor of the US Naval Observatory in Washington DC. Noting that lunar craters possess floors that usually lie below rather than above the level of the adjacent terrain, Gilbert rejected a volcanic origin, saying instead that they were the result of impacts.

Gilbert further proposed that arcuate chains of mountains on the periphery of the 'circular maria' are the walls of vast craters. As evidence, he cited what he described as the *sculpture* produced by the fall of ejecta from the formation of the feature that contains the Sea of Rains. He announced his observations in *The Moon's Face: A Study of the Origin of its*

Features, which was a paper presented orally to the Philosophical Society of Washington later that year to mark his retirement as its president.

When critics asked why there had been no similar impacts on Earth, Gilbert replied that there would have been such occurrences but the craters would rapidly erode.

Gilbert's paper was formally published in the *Bulletin* of the USGS in 1893 but because this wasn't on the reading list of astronomers his intrusion into their bailiwick passed unnoticed.

It was not until R. B. Baldwin made an analysis of bomb craters in the Second World War that the impact origin of lunar craters began to make real headway. As a

IMBRIUM SCULPTURE

When G. K. Gilbert observed the Moon in a powerful telescope in 1892, his geologist's eye was drawn to a pattern of long narrow streaks that radiated out from the chains of mountains that bound the Sea of Rains. He postulated that the streaks were produced by ejecta hurled out on very shallow trajectories during the formation of the vast crater whose rim is marked by the arcuate chains of mountains.

Above is Gilbert's sketch of what he called *sculpture* (*The Moon's Face: A Study of the Origin of its Features*, Philosophical Society of Washington, 1892), together with an image of the Moon that indicates the area contained in the picture adjacent, which in turn supplies context for the view on the facing page in which such grooves are evident.

(Woods using NASA/GSFC/Arizona State University imagery).

businessman trained in physics, he developed an interest in the Moon in 1941 during a visit to a planetarium when, in viewing the photographs on display, he independently noticed the sculpture. Upon finding no explanation in the literature (because he didn't happen across the paper by Gilbert) Baldwin decided to undertake his own study.

In articles published in the magazine *Popular Astronomy* in 1942 and 1943 Baldwin suggested that the process which excavated the cavity in which the Sea of Rains now resides had ejected some of the crustal material on shallow trajectories and the rest on high ballistic arcs. The nearest sculpture consisted of grooves scraped by material moving horizontally, and the chains of craters farther out were made by plunging debris.

Baldwin published in a popular outlet because he was refused by professional journals – as if the Moon wasn't an object worthy of study. In his book *The Face of the Moon*, published by the University of Chicago in 1949, Baldwin drew on his professional analysis of bomb craters. He showed that the greater the deceleration upon impact, the greater was the energy that was released. He then argued that the weak lunar gravity would permit an explosion to toss ejecta to a greater distance but it wouldn't actually make the crater any larger. He logarithmically plotted the relationship between the diameters and depths of explosive craters on Earth, those terrestrial craters which were recognised to be cosmic impacts, and the lunar craters whose walls had not yet slumped to distort the ratio that he used. This plot revealed a clear trend.

Significantly, Baldwin identified sculpture that was associated with other 'circular maria'. He also realised that the mountains peripheral to the Sea of Serenity must have formed prior to the impact that created those peripheral to the Sea of Rains.

Like Gilbert, Baldwin presumed all the maria were made simultaneously. The focus of attention was the impact that made the cavity that accommodates the Sea of Rains, because this was presumed to have been the greatest impact in lunar history. Gilbert had imagined this event 'splashing out' a fluid ejecta that pooled in widely distributed low-lying areas to produce the various maria, but Baldwin knew there had been a significant interval between the formation of the cavity and its being filled in by the Sea of Rains.

Baldwin proposed a scheme by which the creation of the Sea of Rains could release a pulse of extremely fluid lava which burst out through the containing walls to spread across the surface and pool in successively adjacent cavities. Fluid lavas on Earth had left vast flows such as the Columbia River Plateau in America and the Deccan Traps in India.

The important point identified by Baldwin was that the large circular cavities were made by a succession of impacts, with a significant period of time elapsing before the cavities were occupied by the maria.

Astronomers weren't impressed by Baldwin's arguments and continued to associate the creation of the maria with the formation of the cavities in which they reside.

Harold Urey and Gerard Kuiper

In the post-war years, two distinguished scientists became figureheads for rival theories of how the Moon came to be.

Harold Urey won the Nobel Prize for chemistry in 1934 at the age of 41 for discovering deuterium, the 'heavy' isotope

RIGHT Baldwin's book.

The FACE of the MOON

BY RALPH B. BALDWIN

An answer for the age-old question: How and when did the moon acquire its peculiar craters, rays, lava flows, and mountains?

of hydrogen. Later, when at the University of Chicago, he performed experiments to study Earth's early atmosphere and how life might have originated.

After reading Baldwin's book, Urey developed an interest in the Moon. But being a chemist he was more interested in its composition than its surface features. Urey accepted that craters were made by impacts, but he rejected Baldwin's conclusion that a significant interval had elapsed between the formation of the gigantic circular cavities and the formation of the maria by a vast pulse of volcanism. Urey agreed with Gilbert that the maria were formed simultaneously by a splash of impact melt.

In *The Planets: Their Origin and Development*, which was based on lectures he gave at Yale University and was published in 1952, Urey proposed that the Moon condensed from the solar nebula independently of Earth and was later captured. He also said that the interior of the Moon wasn't thermally differentiated into layers, it was 'pristine' material from the nebula. This 'cold Moon' hypothesis implied that volcanism couldn't have played a role in forming the surface features.

Born in 1905, the son of a tailor in Holland, Gerard Kuiper developed an early interest in astronomy. After graduating from the University of Leiden in 1927 and gaining his doctorate in 1933, he relocated to the Lick Observatory in California and spent some time at the Harvard College Observatory before settling at the Yerkes Observatory of the University of Chicago. In 1954 Kuiper proposed that Earth and the Moon had formed in a common envelope within the solar nebula and have always been gravitationally bound.

In an age in which photography was the norm, Kuiper placed a binocular eyepiece on the 208cm reflector of the McDonald Observatory in Texas to discern details of the lunar surface in moments of exceptional clarity that would have been blurred in pictures.

In 1954, in his first paper on the topic, Kuiper said that early radiogenic heating in a body of the Moon's size would have caused sufficient melting for dense minerals to sink to create a core and for lightweight

ABOVE **Harold Clayton Urey.** *(Bettmann/Corbis)*

minerals to rise and form a crust. This thermal differentiation would become known as the 'hot Moon' hypothesis. Because volcanism is a way for heat to escape from the interior, Kuiper argued that the maria were produced by magma upwelling at various times from deep fractures in the floors of the cavities excavated by major impacts.

As these hypotheses were mutually exclusive, Urey and Kuiper became intense rivals.

LEFT **Gerard Peter Kuiper.** *(McDonald Observatory)*

A terrestrial impact

Meanwhile, an insight into the impact process was gained by studying a crater on Earth.

In 1903 mining engineer D. M. Barringer began to drill through the floor of a bowl-like cavity known as Coon Butte in the Arizona desert, between the towns of Flagstaff and Winslow. It had a diameter of 1.2km, a rim that stood 45m above the plain, and a floor 200m below the rim. It was commonly believed to have been created some 50,000 years ago when hot magma caused underground ice to flash to steam and blast a hole in the rock above. But Barringer suspected it was an impact crater and wanted to find out whether there was an iron meteorite beneath the cavity. He found nothing of commercial value.

Astrophysicist E. J. Öpik realised in 1916 that the impact of a large meteorite was so violent that the projectile would be vaporised in an instant. But he published in an Estonian journal and his insight passed unnoticed. A. C.

RIGHT The Canyon Diablo meteorites are scattered around Meteor Crater. This is the largest recovered piece. *(Wikipedia)*

Gifford independently had the same insight in 1924. He published in a New Zealand journal which had a broader readership.

Öpik and Gifford had found that a high-speed impact always creates a circular crater, because although momentum is a vector quantity, energy isn't. As the projectile hits the ground it explodes, releasing its kinetic energy in a symmetrical manner and forming a circular crater. Furthermore, the crater is always much larger than the projectile.

If the Coon Butte crater was indeed an impact, then the only relics of the projectile are the Canyon Diablo meteorites that litter the surrounding desert, the largest of which has a mass of 639kg.

When Gene Shoemaker joined the USGS in 1948 he already had an interest in the Moon. His review of the literature found Gilbert's paper and Baldwin's book, both arguing an impact origin for lunar craters.

In 1955 Shoemaker investigated two craters about 100m wide which were made by underground nuclear explosions in Nevada. His motivation was to find out how such explosions shock and disperse rock. He was impressed by the resemblance of the holes to lunar craters. In 1957 he began a study of Coon Butte. He had already done the field work for a doctoral thesis on another theme but had yet to write this up. After hearing Shoemaker present a seminar about his investigation of Coon Butte, his advisor at Princeton, Harry Hess, suggested he use that as the basis of his thesis. Shoemaker undertook some further field work, wrote it up, and submitted his thesis in 1959.

The fact that the Coon Butte crater was recent (geologically speaking) and in a desert made the way in which it was excavated readily apparent to the experienced eye. In particular, the strike had not penetrated the surface and pushed the rock aside, as Gilbert had imagined an impact ought to do; the process, as Öpik and Gifford had inferred, was explosive. Significantly, Shoemaker found that *two* shock waves were involved – one vaporised the projectile whilst the other propagated downward into the 'target rock', compressing this so thoroughly that the rock reacted precisely as if an explosion had occurred beneath it (like in an underground nuclear detonation). Shoemaker methodically traced how the blanket of ejecta produced by the explosion flipped the stratigraphy into an inverted sequence – as if by a circular 'hinge' at the rim of the crater – thereby making a hole that was much wider than the projectile.

In 1960 L. R. Stieff of the USGS sought NASA funding for an investigation of lunar geology. The space agency deliberated. In 1953 Loring Coes Jr, a chemist at the Norton Company, had produced a new, very dense mineral using a hydraulic press to squeeze quartz. This *shocked quartz* was named *coesite*. In 1956 H. H. Nininger, who specialised in the scientific study of meteorites, had suggested searching for coesite at Coon Butte

LEFT Eugene Merle Shoemaker. *(USGS)*

BELOW After a field investigation of Meteor Crater, Shoemaker issued this geological interpretation showing how the impact process deposited excavated material. *(USGS)*

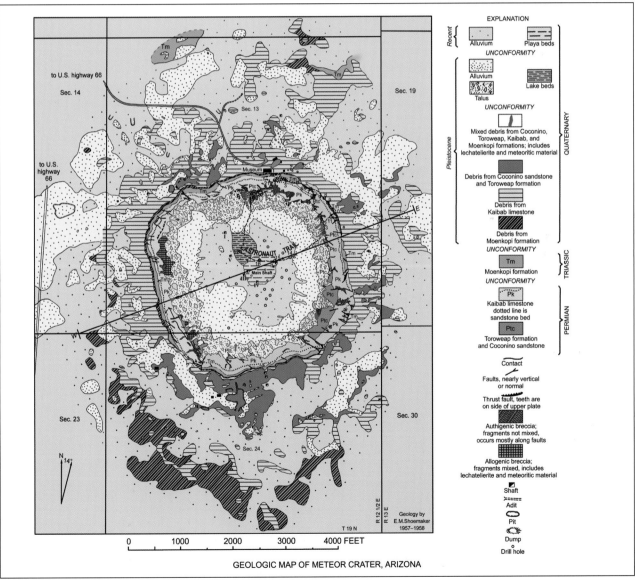

GEOLOGIC MAP OF METEOR CRATER, ARIZONA

but this wasn't done. In 1960 Stieff obtained samples of rock taken from Coon Butte that were in the archive of the Smithsonian Institution in Washington DC, in order that he could tell NASA that the USGS was already at work on craters.

After acquiring his doctorate in geology from the University of Chicago in 1948, Chinese-born Ed Chao joined the USGS. When Stieff gave Chao the Coon Butte samples, Chao identified coesite using an X-ray diffraction analysis. By *proving* Coon Butte to have an impact origin, this research lent strong support to the hypothesis that the lunar craters were the sites of even larger impacts.

After Chao wrote up the scientific paper, the USGS added Shoemaker and his assistant Beth Madsen as co-authors to imply that it had a team of specialists at work. After the paper appeared in the premier journal *Science* in July 1960 NASA released funding to the

USGS, which immediately set up a study group in Menlo Park, California, with Shoemaker in charge. In 1961 this became the Branch of Astrogeology. The following year most of the team moved to Flagstaff, Arizona, where it has remained to this day.

Stratigraphic mapping

Meanwhile, in 1957 the National Academy of Sciences awarded Gerard Kuiper funding to start work on a new lunar atlas. Additional money was provided by the Air Force. The resulting *Photographic Lunar Atlas* was issued in 1960, with pictures printed on a scale at which the lunar disc would be 2.5m across.

Naturally the USGS wanted to map the Moon geologically. The key first step was to identify the various distinct *geological units* in terms of their textures, to delineate their outlines

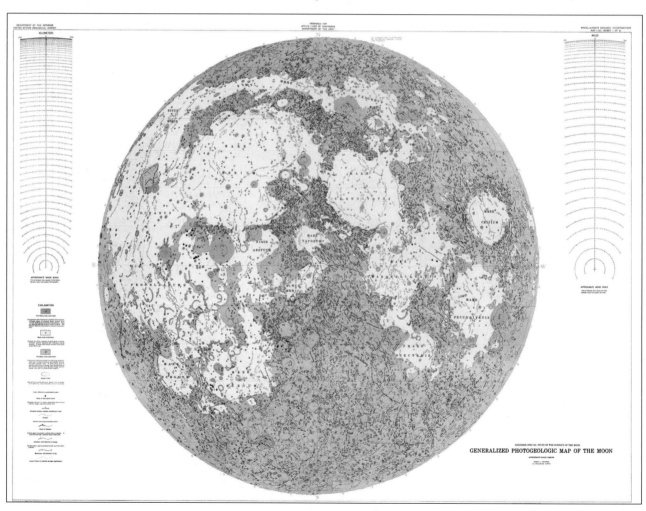

THE APENNINE BENCH FORMATION

In 1961, while mapping relationships between what he defined as 'pre-maria', 'maria', and 'post-maria' units, Robert Hackman of the USGS noted a patch of light-toned material between the crater Archimedes and the arcuate Apennine Mountains. Ejecta from Archimedes was evident on this patch, and it had been encroached upon by the dark material of the Sea of Rains. By the law of superposition, in which older units are buried by later ones, the sequence was clear: the light-toned material was the floor of the enormous cavity made by the impact that produced the mountain chain; later a smaller impact had created Archimedes; and when the Sea of Rains later encroached it had submerged all of the Archimedes ejecta except for that on the more elevated 'bench'. This therefore became known as the Apennine Bench Formation.

It was initially presumed that the bench was impact-melt produced during the creation of the Imbrium basin. However, although several rock samples that were gathered by Apollo 15 from the Apennines a little farther to the north may well have come from this formation, they didn't have the inclusions which could be expected to have become incorporated into a shock-melt. Having analysed the chemistry of these samples, Paul Spudis of NASA's Lunar and Planetary Institute concluded that they were an *aluminous* lava whose extrusion predated the dark lavas of the Sea of Rains. With so much of the crust having been cleared away during the making of the basin, the force of isostasy would have caused the basin floor to rise, cracking the remaining crust sufficiently to allow shallow lavas to ooze out to veneer its floor.

By counting craters of all sizes and making assumptions about the population of impactors, geologists have developed a scheme for dating a surface. Such an analysis found there had been little activity in this part of the Imbrium basin for 500 million years, then a series of extrusions of basalt from deeper reservoirs produced the Sea of Rains.

By interpreting this region in terms of the law of superposition, Hackman wrestled the study of lunar history away from the astronomers and set the scene for a revolution in our understanding of the Moon.

(Woods using NASA/GSFC/Arizona State University imagery).

ABOVE The main dome of the Lowell Observatory built by Percival Lowell on Mars Hill in the 1890s. *(Lowell Observatory)*

BELOW The 61cm refracting telescope of the Lowell Observatory. *(www.nickcook.net)*

on a *base map*, and to use the principle of superposition to identify the sequence of their deposition. This was an example of stratigraphic analysis with the objective of deriving insight into the history of the surface.

The pioneering work on this was undertaken by Robert Hackman of the USGS Photogeology Branch in Washington DC, who demonstrated in 1960 that stratigraphic analysis could indeed be applied to the Moon.

When issued in 1961, Hackman's map of what he identified as the 'pre-maria', 'maria', and 'post-maria' surface units marked a significant departure from the manner in which astronomers made their maps. The superposition relationships suggested to Hackman the maria were volcanic flows rather than splashes of impact melt.

Hackman drew particular attention to a patch of light-toned material between the crater Archimedes and the arcuate Apennine Mountains which partially enclose the Sea of Rains. The sequence was clear. The light-toned material was the floor of the cavity produced by the impact that formed the arc of mountains. This had been struck to create Archimedes some time later, as revealed by the impact's ejecta over the light-toned surface. The mare material which filled in the cavity had encroached, but because the light-toned patch was more elevated it had not been submerged by this lava.

Gene Shoemaker had independently made a superposition study of another section of the Moon to demonstrate the stratigraphy technique.

During a visit to a bookstore he had happened upon a photograph of the area around the crater Copernicus taken by Francis Pease at Mount Wilson in 1919. It was of sufficient clarity to distinguish craters as small as 1km across, so he had it enlarged and set about mapping.

Whereas Hackman had mapped only three units, Shoemaker used seven. In March 1960 he presented a paper showing that whereas much of the material excavated by Copernicus had been 'hinged' to create the rim and adjacent ejecta blanket, some material had been hurled farther to produce chains of small secondary craters. These secondaries were less energetic because, for the ejecta to have fallen back it could not have exceeded the escape velocity of the Moon, which is an order of magnitude slower than the typical speed of a celestial impact.

Shoemaker's stratigraphic mapping not only confirmed Copernicus to be an impact crater, it also refuted the assertion by those who favoured the volcanic origin of lunar craters that the chains of small craters close to large craters were associated with crustal fractures.

Shoemaker and Hackman presented a joint paper at the International Astronomical Union's Symposium in December 1960.

Meanwhile, a new round of ordinary surface mapping was also underway.

In 1959 the Air Force Aeronautical Chart and Information Center in St Louis, Missouri, began to employ airbrushing to represent topography in Lunar Astronautical Charts on a scale of 1:1,000,000.

In 1961 Kuiper convinced the Air Force that there was merit in using visual observations when compiling these charts, because the eye is able to resolve finer detail in brief moments of clarity than can be recorded during a photographic exposure. The pictures would provide the positional basis for mapping and the visual observations would supply the detail. To discern the subtle topography these observations were to be made just after lunar sunrise or just before sunset.

Two large refracting telescopes were made routinely available for this activity – the 61cm of the Lowell Observatory in Flagstaff and the 92cm of the Lick Observatory in California.

Lunar basin structures

In February 1960 the University of Arizona in Tucson recruited Kuiper as the first director of its Lunar and Planetary Laboratory.

When William Hartmann was hired by LPL in 1961 with a background in physics, geology and astronomy he was assigned to the team which was working to improve the *Photographic Lunar Atlas*. Although the best photographs of the limb regions were blurred by atmospheric turbulence, it was possible to project them onto a white globe and then photograph that

in order to eliminate the foreshortening seen by a terrestrial observer and thereby obtain a new perspective of the limb regions.

This *Rectified Lunar Atlas* was published in 1963. One major discovery was the existence of systems of concentric rings. These hadn't been recognised from Earth due to limb foreshortening, but when viewed from an 'overhead' perspective they stood out clearly. The most spectacular such feature had at its centre a small dark patch that could be glimpsed only at times of favourable libration and had been named the Oriental Sea by astronomers.

On realising that the multiple-ring structures were a distinct class of feature, Hartmann coined the term *basin*. He wrote up the discovery with Kuiper and published it in-house at LPL in June 1962. Soon, similar patterns were identified in degraded states around a dozen of the 'circular maria'. This new insight revealed the true violence of a basin-forming impact. Such an event excavated a cavity at the point of impact and the shock made one or more concentric rings of mountains composed of individually faulted blocks which faced steep slopes inwards. Beyond these rings, a *base surge* deposited a blanket of fluidised material, grooves were sculpted by blocks ejected laterally, and secondary craters were made by

BELOW William Hartmann photographing a picture of the near side of the Moon projected onto a half-sphere in order to eliminate the foreshortening that impairs a terrestrial view of the limb region. *(Lunar and Planetary Laboratory of the University of Arizona)*

the debris that fell on steep trajectories – with all of this occurring in an instant.

After a considerable interval, perhaps many millions of years, lava erupted from deep fractures in the cavity to flood it, often sufficiently to submerge the inner rings. The fact that a basin was distinct from the mare which formed later was highlighted by the discovery of a number of large craters that possessed concentric rings and hadn't been filled with mare. It also became evident that many well-known sizeable craters are probably

THE ORIENTALE BASIN

As photographic techniques improved and professional observatories published their lunar atlases, amateur astronomers devoted their attention to mapping the regions which are viewed in highly foreshortened form at favourable librations.

There are mountains on the limb nearby the Sea of Moisture and in his 1906 book *Der Mond* (*The Moon*) the German astronomer Julius Franz described a small dark plain beyond them. At that time the leading limb was the eastern one because it faced the eastern horizon for an observer, so he named this the Eastern Sea. It was independently discovered a generation later by Percy Wilkins and Patrick Moore in England.

When William Hartmann was recruited by the Lunar and Planetary Laboratory of the University of Arizona in 1961, he joined the team developing a technique designed to improve knowledge of the limb regions. The best pictures of the limbs were projected onto a white globe and photographed from the side. This eliminated the foreshortening seen by a terrestrial observer. It revealed the presence of a system of concentric rings centred on the Eastern Sea. When observed conventionally they were simply a mass of mountains but when viewed from 'overhead' their true form stood out clearly. However, because most of the structure was on the far side only part of it was on display in this manner.

This multiple-ring structure was named the Orientale basin for consistency with the Eastern Sea. Later in that year, however, the International Astronomical Union flipped the convention so that the eastern limb was the one over which the Sun would rise for a person on the lunar surface, rather than the one which faced the eastern horizon of a terrestrial viewer. Consequently the Orientale basin which contains the Eastern Sea is now on the western limb!

After the Lunar Orbiter project had fulfilled its obligation to Apollo by reconnoitring a number of potential landing sites in the equatorial zone, the fourth mission was assigned a general mapping task. On 8 May 1967 it entered an orbit ranging between 2,706 and 6,114km with a 12hr period that was inclined at 85.5° to the equator to enable the ground track to remain over the migrating terminator for optimal observation of surface topography.

The phase of the Moon was 'new' on 9 May and would be 'full' on 23 May. The mapping mission began on 11 May with a repetitive schedule. On each perilune passage the vehicle was to take a series of pictures either side of the equator. On alternating revolutions it was to take pictures at 72° from 3,500km, viewing poleward from 50° latitude. At apolune on every fourth orbit it was to take a single picture of the far side. The system was programmed to scan and transmit the film incrementally while near apolune.

On 20 May the drive of the scanner began to misbehave but it was decided to push on in the hope of being able to extend the coverage over the western limb and photograph the Orientale basin as planned.

The resolution depended on altitude but at perilune it was 60m, which was better than the 300m that was possible for a terrestrial observer employing a large telescope in excellent atmospheric conditions. The results revealed new geological insights of the polar and limb regions.

A picture from an altitude of 2,723km showed the Orientale basin and at last the scientists were able to see the majority of this feature. The mare material hadn't completely flooded its central ring, so the vast scale of the impact structure was open to inspection.

not primary impacts but secondaries from basin-forming events.

In just a few years, therefore, an examination of the Moon by geologists applying their mapping methods had yielded insights into the history of the lunar surface which had eluded astronomers for centuries because the latter had been delving into ever finer details without recognising the 'big picture'.

Nevertheless, as the Space Age dawned Urey and Kuiper were at loggerheads concerning the origin and history of the Moon.

(NASA)

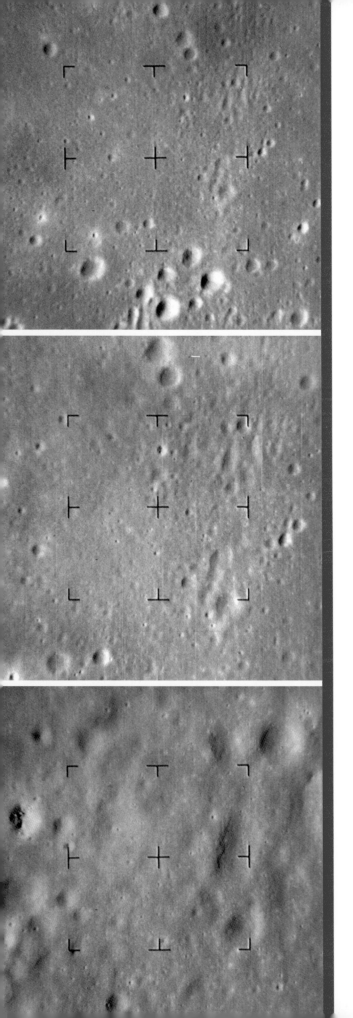

Chapter Six

Early lunar missions

This chapter discusses the spacecraft that were sent to the Moon at the dawn of the Space Age in order to reveal its hidden far side and to find out what the surface was like on a scale closer than could be achieved by terrestrial telescopes.

OPPOSITE An Atlas-Agena launches Ranger 7, and a nested set of images transmitted by the spacecraft as it dived to destruction on the lunar surface. The bottom frame is the final one transmitted by the 'A' camera. *(NASA)*

In the early years of the Space Age the Soviet Union reaped spectacular successes. It all began with the launch of the world's first artificial satellite in October 1957, with its name Sputnik meaning 'Fellow Traveller'.

In 1959 they began to shoot for the Moon. In January the first Lunik was intended to hit the Moon but it missed by 5,000km and became the first human artefact to pass into an independent orbit around the Sun, which was an achievement in its own right. A second probe in September did strike the Moon in the Sea of Rains between the craters Archimedes and Autolycus.

These probes had instruments which could measure magnetic fields in space. It transpired that the Moon doesn't possess one. Earth's magnetic field is a dynamo generated by electric currents within the core of molten iron. The absence of a dipole field suggested that if the Moon ever had a molten core, either this had since solidified or the Moon's axial rotation was too slow to generate electric currents.

By missing the Moon the first probe, which was subsequently nicknamed Mechta ('Dream'),

confirmed the presence of a magnetic field in interplanetary space associated with a plasma of charged particles streaming away from the Sun; a phenomenon known as the *solar wind*.

The far side revealed

In October 1959, Lunik 3 was launched into an elongated orbit of Earth that looped around the Moon, where a camera snapped a number of pictures. After the film had been developed, the pictures were scanned and transmitted to Earth.

Since the Moon's axial spin is synchronised with the period of its orbit around Earth, its rear hemisphere was a mystery. If it resembled the visible side, then it would be dominated by dark plains.

At the time the photography was carried out, the Moon was a waxing crescent as seen from Earth. This meant most of the far side was illuminated. Although the quality of the pictures was poor, they gave a new perspective of the maria on the familiar crescent that are highly foreshortened to terrestrial observers and also

BELOW A replica of Lunik 1 exhibited in the *Kosmos* pavilion of the Exhibition of Achievements of National Economy of the USSR. *(RAI Novosti/ Alexander Mokletsov)*

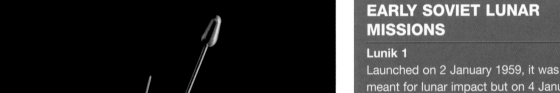

EARLY SOVIET LUNAR MISSIONS

Lunik 1
Launched on 2 January 1959, it was meant for lunar impact but on 4 January it missed the Moon by some 5,000km and entered heliocentric orbit.

Lunik 2
Launched on 12 September 1959, it impacted in the Sea of Rains on 14 September, close to the meridian and about 30°N, taking measurements of the space environment during the approach.

Lunik 3
Launched on 4 October 1959, it was inserted into an elliptical Earth orbit which passed around the back of the Moon to provide an opportunity to photograph the far side on 7 October. As it flew back towards Earth it transmitted the pictures.

revealed that there are only a few small mare plains on the far side.

This successful mission marked the end of the first phase of the Soviet investigation of the Moon. *(See 'First view of the far side of the Moon' overleaf)*

The surface in close-up

The Ranger project was initiated in December 1959 as NASA's flagship for exploring the Moon. It was managed by the Jet Propulsion Laboratory (JPL) of the California Institute of Technology in Pasadena.

The initial idea was for a vehicle to make a near-vertical dive towards the Moon. A television camera would be activated 4,000km from the Moon, to transmit pictures to Earth until a capsule was released that would fire a rocket to slow its rate of fall sufficiently to place a small package of instruments on the surface.

Having developed a television system for meteorological satellites, the Astro-Electronic Division of the Radio Corporation of America in Hightstown, New Jersey, supplied a slow-scan vidicon imaging tube and associated electronics for the Ranger spacecraft. JPL built the optical element, which was an f/6 telescope with a focal length of 1m. The image scanned off the vidicon tube would have 200 'lines'.

The plan was for the camera to provide 100 pictures, with the transmission being terminated when the release of the capsule at an altitude of about 24km disturbed the high-gain antenna sufficiently for it to lose its lock on Earth.

In October 1961 Gerard Kuiper, Harold Urey, and Gene Shoemaker were named as the experimenters who would receive and interpret the pictures.

In excellent seeing conditions the best telescope on Earth has a lunar surface resolution of 300m. An image taken by the Ranger camera at an altitude of 50km was expected to provide a resolution 100 times better. The results were eagerly awaited, but as events transpired none of the three craft equipped in this manner were able to return any pictures.

In December 1962, with its best result being a dead probe hitting the Moon, NASA redefined the project's goals: the next vehicles would have only a TV system, which would

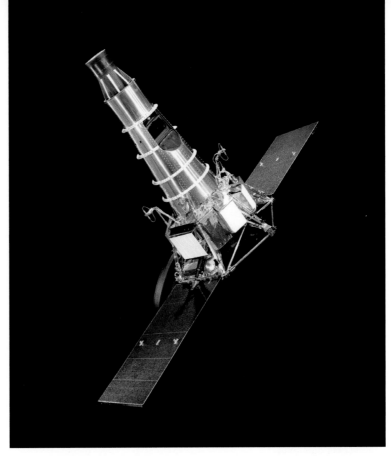

be more sophisticated than the original, and the sole objective would be to obtain close-up pictures of the lunar surface in order to assess whether it was likely to support the weight of a spacecraft. The initial pictures of a mission would be taken from a range that would match the best obtainable by telescopes. Ideally, the spacecraft would make a very steep dive so that successive images would overlap, especially towards the end of the sequence. Imaging would continue through to impact. Watching the final moments of the descent in *real time* would be a stunning experience. The subsequent analysis to establish how the character of the lunar surface varied on different scales would be painstaking.

Kuiper was the principal investigator for the revised imaging project, but the science team included Urey, Shoemaker, and Ewen Whitaker, who was a colleague of Kuiper's at the Lunar and Planetary Laboratory of the University of Arizona.

When the first of the new batch, Ranger 6, was launched in February 1964 the team was optimistic, but as the vehicle dived right on target, 15° east of the meridian in the Sea of Tranquillity, it failed to transmit pictures.

An analysis found that the TV system had been crippled by an electrical arc (essentially

ABOVE A replica of the type of Ranger spacecraft whose primary role was to image the lunar surface during a terminal dive. *(NASA/ JPL-Caltech)*

FIRST VIEW OF THE FAR SIDE OF THE MOON

FAR RIGHT A replica
of Lunik 3. The Sun
sensors are on the
upper and lower
cones, small attitude
control thrusters
are on the bottom
cone, and the camera
window is at the
top. Solar cells are
wrapped around
the cylindrical body,
which is about 1m in
diameter. *(NASA)*

When Lunik 3 was launched on 4 October
1959 it was inserted into a highly elliptical orbit
of Earth which would permit it to observe the
far side of the Moon. Its predecessors, sent
to impact the Moon, were battery powered,
but to sustain the duration required for this
photographic mission the vehicle had a ring of
solar transducer cells around its middle.

The trajectory crossed the radius of the
Moon's orbit at a point 6,200km south of
that orb. After the point of closest approach
on 6 October, the Moon's gravity deflected
the trajectory northward. Whilst in coasting
flight, the spacecraft was spin-stabilised. On
approaching the Earth–Moon plane beyond the
Moon, gas jets were fired to cancel the spin,
a Sun sensor locked on, and the vehicle rolled
around its major axis until a sensor indicated
that the optical system was viewing the Moon.

The camera possessed a 200mm f/5.6
lens and a 500mm f/9.6 narrow-angle lens,
with both pointing in the same direction. At a
range of 65,200km on 7 October, the system
simultaneously exposed pairs of images onto
a 35mm film which had a 'slow' rating in order
to resist its being 'fogged' by the radiation in
space. After 40min of photography the vehicle
resumed its spin for stability. After the 480,000km
apogee on 10 October it headed back in for the
47,500km perigee on 18 October.

The film was wet-developed, fixed, dried,
then scanned using a constant-brightness light
beam which was detected by a photoelectric
multiplier whose output provided an analogue
signal. There were two transmission rates: a
slow rate for when far from Earth and a higher
rate for use at perigee. Radio contact was finally
lost on 22 October.

On Earth, the signal was recorded on
magnetic tape for further processing. The raw
pictures were marred by bands of interference
'noise' which had to be 'removed'. Then the
image contrast was stretched in order to
emphasise detail. In the wide lens the disc of
the Moon spanned 10mm, and in the narrow
lens it was 25mm.

The phase of the Moon was 'new' on 2
October and 'first quarter' on 9 October.
Consequently, when the pictures were taken
on 7 October a portion of the near side was
illuminated and the Sea of Crises provided a
sense of perspective. The remainder of the
illuminated zone was 70% of the hitherto
unobserved far side.

Since the Sun was directly behind the

BELOW One of the
better pictures from
Lunik 3 showing a
view around the limb
onto the far side of the
Moon. To a terrestrial
observer it was a
crescent, hence the
Sea of Crises is on
the left of this view.
*(Academy of Sciences
of the USSR)*

camera, the absence of any shadows made it impossible to discern the topography. Nevertheless, the pictures did show there to be few dark features, indicating the hemispheres that face towards and away from Earth to be different. Dark plains cover 30% of the near side but only 2% of the far side; a total of 16%.

The paucity of maria on the limb and the fact that those which were present were patchy rather than major plains had prompted speculation that there would be fewer of them on the far side, but their virtual absence was a surprise. The fact that the Moon's axial rotation is synchronised with the period of its orbit may well be related to this hemispheric dichotomy.

The Lunik 3 mission was a remarkable success for the first attempt at this difficult task. The results were published in 1960 as *Atlas Obratnoy Storony Luny* (*Atlas of the Far Side of the Moon*) by N. P. Barabashov, A. A. Mikhaylov, and Yu. N. Lipskiy. It showed the far side of the Moon as a disc 35cm in diameter. Ignoring the prerogative of the International Astronomical Union, it assigned names to 500 features of various types. The two most prominent dark patches became the Moscow Sea and Tsiolkovsky, the latter for the Russian pioneer of astronautics Konstantin Tsiolkovsky.

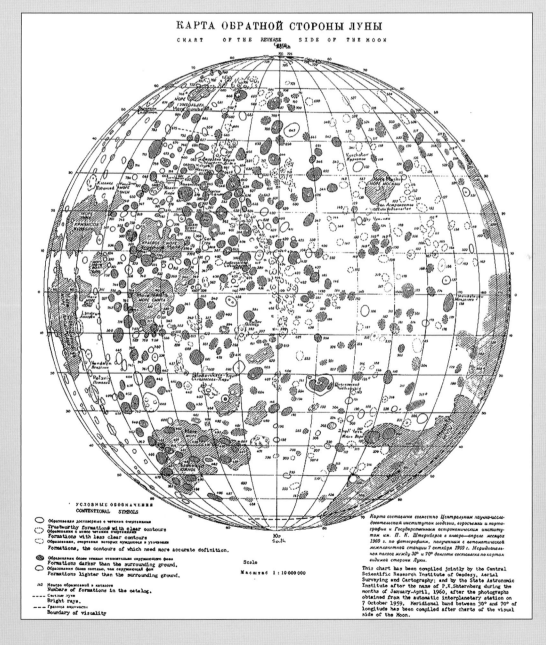

LEFT **The map of the far side of the Moon based on an analysis of the pictures from Lunik 3, published as *Atlas Obratnoy Storony Luny*.** *(Academy of Sciences of the USSR)*

a short circuit) caused by hot gas penetrating an umbilical socket as the Atlas rocket shed its booster section in the ascent from Earth. It was decided to modify the cover plate to prevent this and also to isolate the TV system until it was safely in space.

With Ranger 7 in July 1964, everything was perfect. The target was near the crater Guericke, where the Sea of Clouds was crossed by bright 'rays' from the craters Copernicus and Tycho.

A total of 4,316 pictures were received, the first from an altitude of 2,100km and the last from 500m. At NASA's receiving station at Goldstone in California the spacecraft's signal was first converted into television format and then simultaneously stored onto magnetic tape and displayed on a high-speed monitor that was observed by another camera whose action was synchronised with the incoming frame rate to record each TV frame onto 35mm film. This master film was stored in a vault and a duplicate was made by replaying the tape.

The second film was flown to the Hollywood-Burbank Airport for normal processing by Consolidated Film Industries. Later that same day, prints and slides were driven to JPL to be examined by the experimenters.

The press conference held later that evening was carried 'live' by the national TV networks.

After the JPL Director had introduced the team, Kuiper began the presentation by declaring, 'We've made progress in resolution of lunar detail not by a factor of ten, as hoped would be possible with this flight, nor by a factor of a hundred, which

THE RANGER ADVANCED TELEVISION SYSTEM

The contract for the high resolution television system to be used by the final batch of Rangers was issued to the Radio Corporation of America, the company which supplied the camera for the earlier version of the spacecraft. Its requirements were much more demanding.

It was decided to use a shutter (not a standard feature on a continuous-scan television system) to define a 'frame' on a vidicon tube. The design was finished in September 1961 and had three major subassemblies: a tower superstructure with a thermal shield mounted on top of the hexagonal bus; a central box containing the main electronics; and a battery of six cameras and their individual electronic systems. The television system would have a power supply independent of the bus, and a pair of 60W transmitters. After being activated at a predefined altitude it would continue to send pictures until the vehicle struck the lunar surface.

There were two types of camera. The wide-angle 'A' type had a lens with an aperture ratio of f/1 and a focal length of 25mm, and the narrow-angle 'B' type had an f/2 lens with a focal length of 75mm. There were two 'A' cameras and four 'B' cameras. The vidicons were identical but the entire 11mm square image would be used for a full (F) frame and only the central 3mm square for a partial (P) frame.

One 'A' and one 'B' camera would operate a 5.12sec cycle in which the shutter fired to expose its vidicon and it was read out over an interval of 2.56sec, then erased over the next 2.56sec. They were to operate out of phase so that a frame was taken every 2.56sec.

The other cameras would require 0.2sec to fire the shutter and carry out the readout, and then 0.6sec to erase. The faster cycle time was because a smaller vidicon area was to be scanned. These cameras were to be cycled to provide a frame every 0.2sec in the hope that one would provide a close-up picture immediately prior to impact and reveal the nature of the lunar surface on the finest possible scale.

The cameras were mounted in the tower at angles calculated to provide the overlap needed to establish the relationship between successive frames.

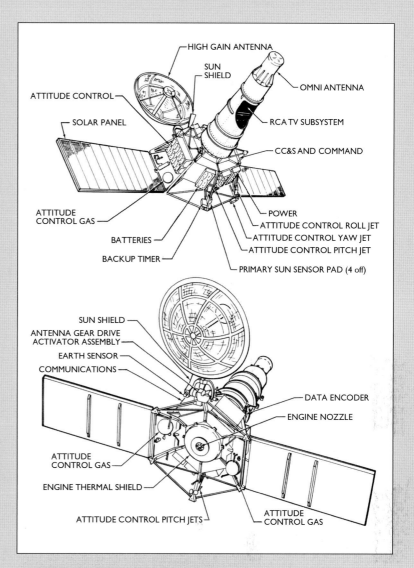

ABOVE Detail of the final batch of Ranger spacecraft that transmitted pictures during a terminal dive. *(NASA/JPL-Caltech)*

BELOW The shutter timing of the six cameras of the Ranger high resolution television system. *(Woods using NASA/JPL-Caltech data)*

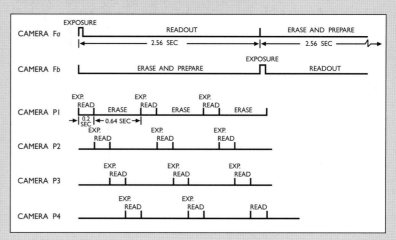

RIGHT A selection of the central portions of nested frames running left to right and top to bottom from the A-camera of Ranger 7 as it made the first-ever documented plunge to the lunar surface. The first frame shown here was taken at an altitude of 1,223km. The final frame from this camera was taken at an altitude of 5.85km and its transmission was not able to be completed. Faster acting narrow-angle cameras recorded detail closer in. *(Harland using NASA/JPL-Caltech imagery)*

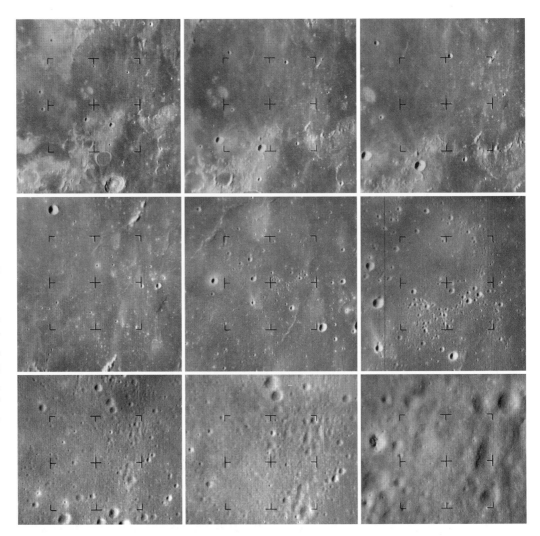

RIGHT A composition of the final (incomplete) frame from the A-camera of Ranger 7, a closer view provided by the panoramic camera aboard Apollo 16 which shows the crater produced by the impact of the unmanned vehicle and a narrow-angle view from the Lunar Reconnaissance Orbiter mission which shows it in greater detail. *(Harland using NASA/JPL-Caltech/ASU/GSFC imagery)*

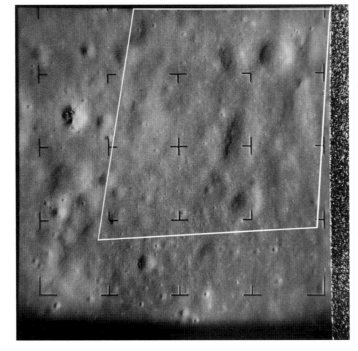

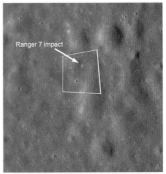

Ranger 7 impact

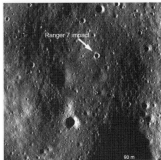

Ranger 7 impact

90 m

would've been already very remarkable, but by a factor of a thousand.' Actually he was a little optimistic, since the resolution of the final frame was more like half a metre.

Urey reported he was 'pleasantly surprised' at the amount of information that could be inferred from the pictures. The surface was cratered all the way down to the limiting resolution of the final frame. The bright 'rays' seemed to be the fall of material on low-energy ballistic trajectories, indicating they were secondaries.

Authors of science fiction had reasoned that because the Moon lacked an atmosphere there would be no erosion, and so artists had shown a rugged landscape. The fact that the surface was gently undulating was a clear indication that the incessant rain of meteoritic material was a potent source of erosion, and Urey coined the term 'gardening' to describe the process by which impacts 'turn over' the surface material.

Kuiper said the pictures supported his belief that the mare plains were lava flows, the surface of which, having been exposed to the vacuum of space, must be 'frothy'. This was an inference based on laboratory experiments in which fluids of various viscosities had been exposed to vacuum. He ventured that when an astronaut walked on the Moon, the experience would be similar to walking on crunchy snow.

In 1955 Thomas Gold, an astronomer with a wide-ranging interest who was then at the Royal Greenwich Observatory in England, had suggested that particles of dust on the lunar surface would become electrically charged by exposure to ultraviolet in solar radiation, and that as the grains of dust repelled one another they would tend to flow and collect in low-lying areas. Tests using powdered cement in a vacuum had indicated a 'fairy castle' structure would form that was full of voids and easily compressible, a result that seemed to be confirmed by bouncing radio waves off the lunar surface. Gold inferred that the maria were deep accumulations of dust, and that any heavy object that landed on the surface would promptly sink from sight.

At the time of the Ranger 7 mission, Gold, now at Cornell University, was not on the team and wasn't at the conference, but he voluntarily explained to reporters that the rims

of the older-looking craters were much softer than the newer-looking ones, and said this was because the slow but remorseless flow of dust grains was smoothing out deformations. Others countered that this softening was caused by the rain of meteorites.

Observing the presence of blocks of rock on the surface which had to be heavy even in the weak lunar gravity, Shoemaker was confident that the surface would bear the weight of a lander.

On 31 August of that year the International Astronomical Union recognised the success of the Ranger 7 mission by naming the impact site the Known Sea.

Whilst the pictures were being used to make an atlas of this region, Kuiper marvelled, 'To have looked at the Moon for so many years, and then to see this ... it's a tremendous experience.'

Ranger 7 gave a flood of data to a community that had been starving for years, fed

ABOVE A plaster model of the lunar surface based upon the most detailed narrow-angle image transmitted by Ranger 7. It reproduces the character of the lunar surface at a resolution almost one thousand times better than was attainable using an excellent telescope in the best of observing conditions. *(USGS)* only by the insistently blurry photographs taken through Earth's unstable atmosphere.

The statistics for primary projectiles are such that there is a small number of really large ones, and then increasing numbers of ever smaller ones. Exposed to such a population the airless lunar surface must gain many small craters and progressively fewer larger ones.

With a presumption of how the population of projectiles has changed over time, the cratering density (as derived from counting) provides an age estimate. If the population is presumed to have remained constant, the sheer number of potential projectiles will have declined but their *size proportions* will have remained the same. It

was evident that large craters would stand apart, but there would be a size of crater at which a given surface would be *saturated* – meaning that such craters would be so numerous that they would occur rim to rim and each new arrival would mask an old crater. This saturation would produce a sharp transition in the *crater curve*. Hence dating by this method can be done only using craters that exceed the saturation size. The saturation size measured from the Ranger 7 pictures was several hundred metres. On smaller scales, the surface was in a *steady state*.

Similar results were obtained by Ranger 8 in February 1965, which imaged a small patch of the Sea of Tranquillity that was free

BELOW A selection of the central portions of nested frames running left to right and top to bottom from the A-camera of Ranger 8. The first frame here was taken at an altitude of 385km and the final frame at 21.8km. The transmission included one additional full frame and one interrupted frame but these are not shown in this composition. *(Harland using NASA/JPL-Caltech imagery)*

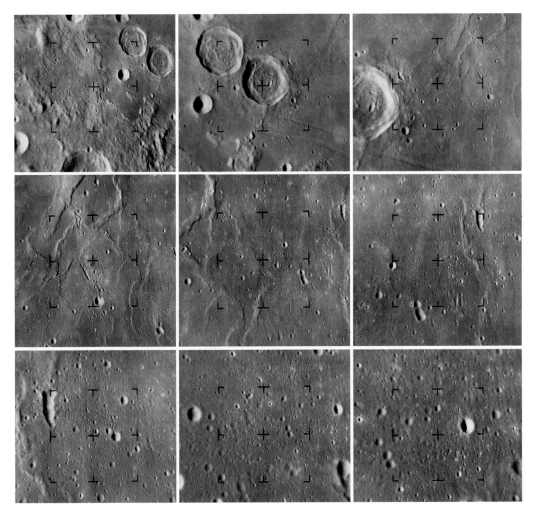

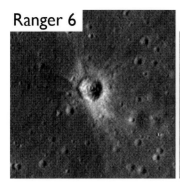
Ranger 6

Ranger 7

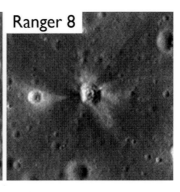
Ranger 8

of bright rays in order to find out whether that was significantly smoother. To reach an easterly longitude, the craft had to fly a more inclined approach. The increased surface coverage was welcomed by the mappers but it resulted in substantial smearing and lack of overlap in the final frames.

Given only the final few frames from the two sites, it was difficult to tell them apart. On a local scale, the two maria were basically the same. Shoemaker observed that this realisation was 'one of the most striking Ranger results'.

With Ranger having satisfied its obligations to the Apollo program, NASA released the final spacecraft to the scientists. Its mission will be explained in a later chapter.

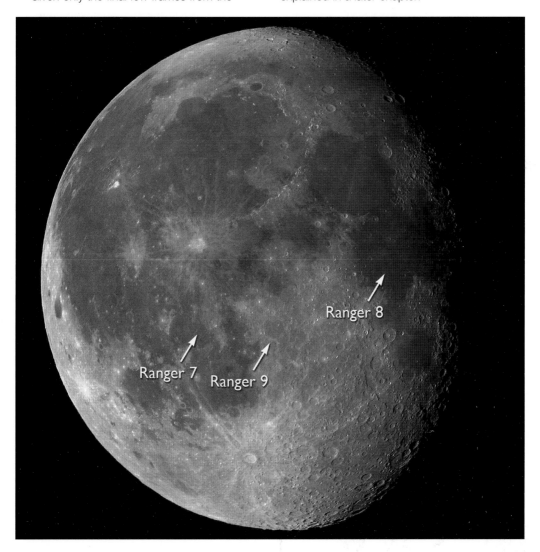

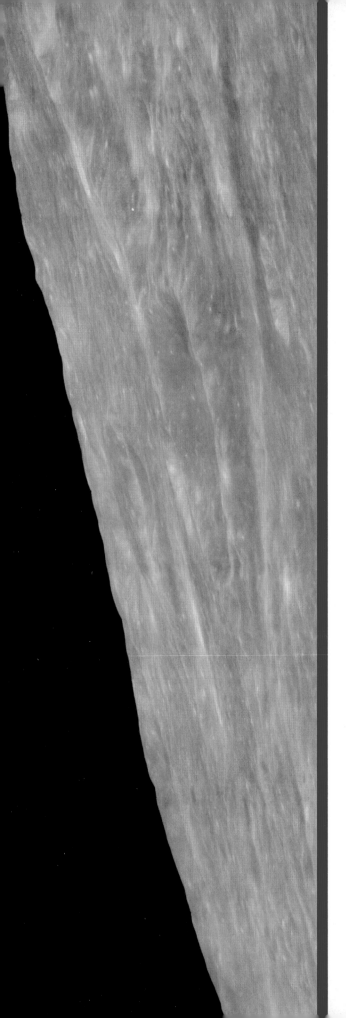

Chapter Seven

Robotic orbiters and landers

This chapter discusses the robotic missions which landed on the Moon to investigate the mechanical and chemical properties of its surface, and others which undertook orbital reconnaissance of potential sites for Apollo astronauts.

OPPOSITE A restoration of the first picture of Earth taken from lunar orbit. The image was taken by Lunar Orbiter 1 on 23 **August 1966.** *(Lunar Orbiter Image Recovery Project, NASA/Ames)*

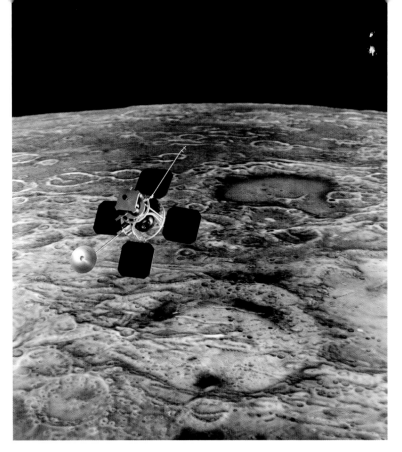

ABOVE **An artist's impression of a Lunar Orbiter spacecraft at work.** *(NASA)*

Lunar Orbiter

In fact, NASA had begun a project to create an orbiter when Ranger was suffering a string of failures, since it was evident that, as a senior agency official put it, for the task of mapping, a single orbiter had more to offer than 'dozens of successful Ranger TV impactors'.

On 30 August 1963 the Langley Research Center in Virginia initiated its first deep-space project, which had the mundane name of Lunar Orbiter. The contract to develop the spacecraft was awarded to Boeing in December.

The nominal mission profile would involve flying to the Moon and the preliminaries in orbit prior to ten days of photography, followed by the development, scanning, and transmission of the film. Hence the spacecraft might have to operate for a month.

The design included a lightweight form of a photographic system that had been created by Eastman Kodak in 1960 for a reconnaissance satellite. It had two lenses that would take wide-angle and narrow-angle frames simultaneously (referred to as the 'medium' and 'high resolution' frames) and interleave them onto a strip of film. The film would be developed and fixed using the 'semi-dry' Bimat process (similar to the Polaroid process) introduced by Kodak in 1961 that eliminated the need to handle 'wet' chemicals in weightlessness.

A key decision for the lunar application was to use Kodak SO-243 fine-grain 'aerial film' in order to obtain the desired resolution. Its exceedingly 'slow' rating of 1.6 ASA meant this film was unlikely to be 'fogged' by the radiation in deep space.

A Ranger spacecraft that transmitted pictures while diving to destruction could provide detailed views of a small area of the lunar surface. What the Apollo planners wanted was similar coverage of large tracts of the near side equatorial zone, to enable them to select a suitable site for the first manned landing. For that they needed a camera in lunar orbit.

Dynamical constraints meant the target for the first Apollo landing had to be close to the lunar equator.

The Sun traverses the lunar sky at a rate of about 12° in 24hr, therefore this is the rate at which the sunrise terminator migrates from east to west.

The candidates for the first Apollo landing had also to be sufficiently far from the eastern limb to provide time for radio tracking prior to initiating the powered descent. And to accommodate two delays of 48hr each on launching the first Apollo landing mission, it was

BELOW **A meeting of the Lunar Orbiter Project Office on 14 December 1965 discusses the 'A' mission.** *(NASA)*

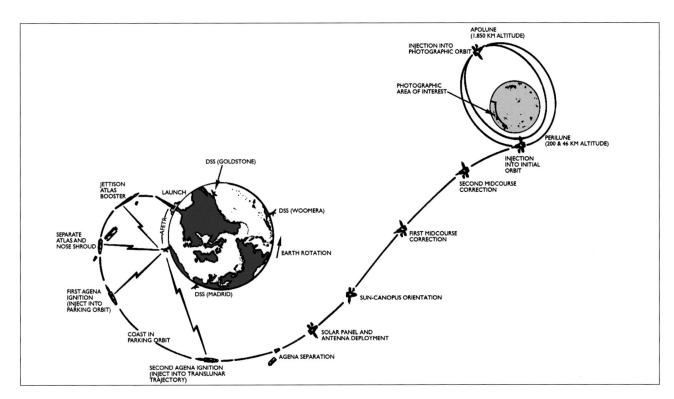

The zone of principal interest was therefore the 'box' lying within 5° of the lunar equator and 45° of the central meridian. It had been calculated that three successful Lunar Orbiter missions would be required to survey this zone but five spacecraft were assembled to allow for failures. Accordingly, three imaging plans were drawn up, labelled 'A', 'B', and 'C'.

Each spacecraft passed beyond the trailing limb of the Moon prior to firing its engine for orbital insertion. The initial orbit had an apolune of about 1,800km and a perilune at about 200km. Another manoeuvre would lower the perilune to 50km at a point close to the sunrise terminator. Because the plane of the spacecraft's orbit would remain fixed relative to the stars, the longitude of perilune would move westward as the Moon travelled around Earth. This permitted each spacecraft to photograph a series of sites along the equatorial zone with the Sun at essentially the same elevation in the sky.

The orbits of the first two spacecraft had the same angle to the equator, but were configured so that the first was optimised to give a vertical perspective of southerly targets and the second for northerly ones. The third mission was set up to undertake follow-up observations of

promising sites right across the zone. It was also to obtain pictures to facilitate stereoscopic analysis to create terrain maps with 3m contours, in order to chart the topography along the line of approach from the east which an Apollo lander would travel during its descent. In addition, some potential target sites were

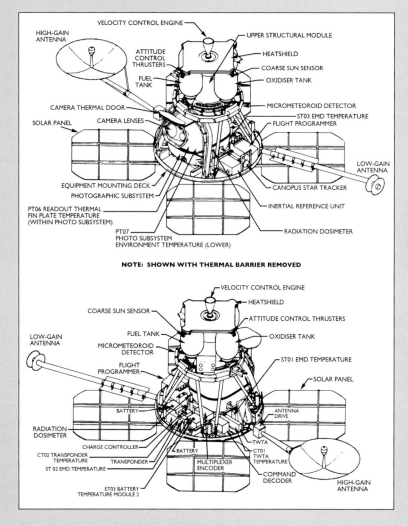

High-gain antenna labels:
- VELOCITY CONTROL ENGINE
- HIGH-GAIN ANTENNA
- ATTITUDE CONTROL THRUSTERS
- UPPER STRUCTURAL MODULE
- HEATSHIELD
- COARSE SUN SENSOR
- FUEL TANK
- OXIDISER TANK
- CAMERA THERMAL DOOR
- MICROMETEOROID DETECTOR
- ST03 EMD TEMPERATURE
- FLIGHT PROGRAMMER
- SOLAR PANEL
- CAMERA LENSES
- LOW-GAIN ANTENNA
- EQUIPMENT MOUNTING DECK
- PHOTOGRAPHIC SUBSYSTEM
- CANOPUS STAR TRACKER
- INERTIAL REFERENCE UNIT
- PT06 READOUT THERMAL FIN PLATE TEMPERATURE (WITHIN PHOTO SUBSYSTEM)
- RADIATION DOSIMETER
- PT07 PHOTO SUBSYSTEM ENVIRONMENT TEMPERATURE (LOWER)

NOTE: SHOWN WITH THERMAL BARRIER REMOVED

Second diagram labels:
- VELOCITY CONTROL ENGINE
- COARSE SUN SENSOR
- HEATSHIELD
- ATTITUDE CONTROL THRUSTERS
- LOW-GAIN ANTENNA
- FUEL TANK
- MICROMETEOROID DETECTOR
- OXIDISER TANK
- FLIGHT PROGRAMMER
- ST01 EMD TEMPERATURE
- SOLAR PANEL
- BATTERY
- ANTENNA DRIVE
- RADIATION DOSIMETER
- CHARGE CONTROLLER
- BATTERY
- TWTA
- CT02 TRANSPONDER TEMPERATURE
- CT01 TWTA TEMPERATURE
- TRANSPONDER
- MULTIPLEXER ENCODER
- ST 02 EMD TEMPERATURE
- COMMAND DECODER
- ET03 BATTERY TEMPERATURE MODULE 2
- HIGH-GAIN ANTENNA

ABOVE Details of the Lunar Orbiter spacecraft. (NASA)

LUNAR ORBITER'S PHOTOGRAPHIC SYSTEM

The 68kg photographic system developed for Lunar Orbiter occupied an ellipsoidal aluminium 'bath tub' about the size of a large suitcase.

The camera was immersed in nitrogen in a pressure vessel, with its lenses viewing through windows of quartz. It was protected by a hinged thermal door that would open for a photographic session. The door was present to eliminate light which might fog the film that was stored in 'loopers' which ran between the various subassemblies. It would also preclude thermal distortion of the optics by maintaining the lenses at a stable temperature.

It had an f/5.6 lens with a 610mm focal length for high resolution (H) narrow-angle frames and an f/2.8 lens with an 80mm focal length for medium resolution (M) wide-angle frames.

The shuttering mechanism offered exposures of 1/25th, 1/50th, or 1/100th second. At the intended perilune of 40km the spacecraft's orbital velocity of 1.6km/sec would cause it to travel 64m during the slowest exposure. To enable the narrow-angle lens to attain the required 1m surface resolution, it was necessary to compensate for this motion. This was to be achieved by using a velocity/height (V/H) sensor which made repeating circular scans

BELOW The two-lens photographic system of the Lunar Orbiter spacecraft. (NASA)

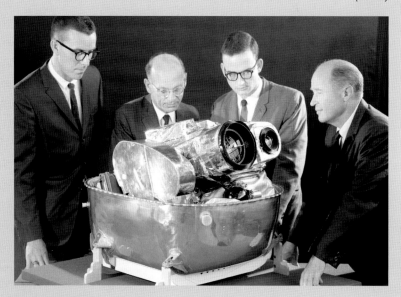

BELOW An Eastman-Kodak engineer works on the Lunar Orbiter photographic system out of its shell. (NASA)

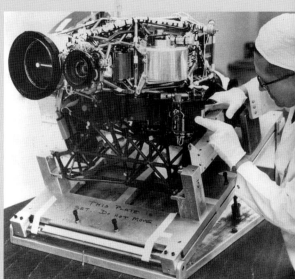

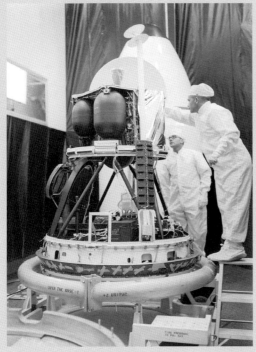

FAR LEFT Assembling a Lunar Orbiter spacecraft, prior to the photographic system being installed in the framework. *(NASA)*

LEFT The photographic system installed in the Lunar Orbiter spacecraft. *(NASA)*

of a portion of the image in the narrow-angle field to calculate the rate and direction of any image motion, and then counter the changing perspective by driving a servo motor to adjust the rate at which the film crossed the focal-plane shutter.

The two shutters were to operate together, so each 'exposure' would place a pair of frames onto the strip of film, with the H frame being the central part of the M frame.

At an altitude of 40km, an M frame would cover an area of 33 × 36km with a resolution of about 7m, and an H frame would cover 4.1 × 16.4km with a resolution of 0.9m. Thus an H frame viewed 5% of the corresponding M frame at 8 times the resolution.

The system could take 1, 4, 8, or 16 exposures in rapid sequence. Contiguous coverage could be obtained by timing the shuttering to obtain a series of overlapping frames at either the M or H scale. A known change in perspective between overlapping frames would facilitate stereoscopic analysis with a given illumination from a single photographic pass.

The specification for the system stated that it must provide at least 194 frame pairs, but Kodak managed to accommodate a film 70m in length which was nominally capable of 212 exposures.

The unperforated film had a thin band of

pre-exposed data running along one edge. This would be read out along with the image. It provided a grey-scale and the calibration charts for determining the resolving power.

Film from the supply reel travelled through the focal plane of the optical system and, after being exposed, was stored in a looper until it was fed to the processor.

The Bimat development process employed a thin gelatine layer of Kodak SO-III film soaked in imbibant type PK-411 solution. This was briefly laminated against the film to develop and fix the negative image in a single step. Once peeled

BELOW An engineering model of the Lunar Orbiter spacecraft. *(NASA)*

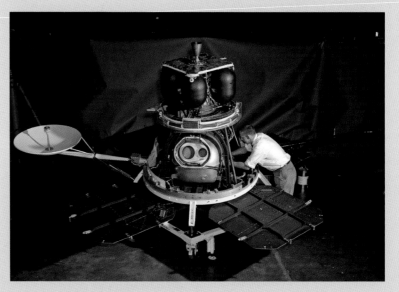

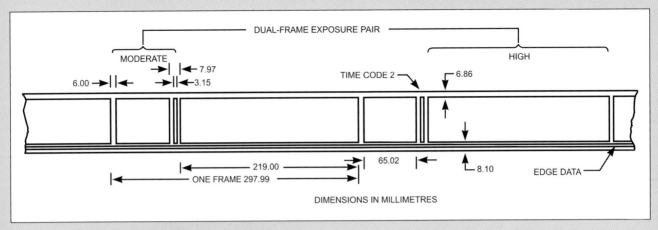

DUAL-FRAME EXPOSURE PAIR

MODERATE

HIGH

6.00

7.97

3.15

TIME CODE 2

6.86

219.00

65.02

8.10

ONE FRAME 297.99

EDGE DATA

DIMENSIONS IN MILLIMETRES

ABOVE Each exposure of the Lunar Orbiter photographic system simultaneously took a square medium resolution frame and a rectangular high resolution frame. They were interleaved on the film. *(NASA)*

RIGHT The overlap of successive exposures of the Lunar Orbiter photographic system could be varied. In some cases the high resolution frames produced contiguous coverage. *(NASA)*

BELOW Details of the Lunar Orbiter photographic system, showing the 'Bimat' process for incrementally developing the film. *(NASA/Woods)*

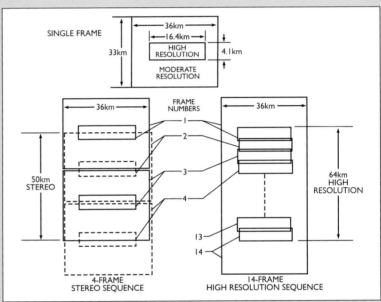

SINGLE FRAME

36km
16.4km
33km
HIGH RESOLUTION
4.1km
MODERATE RESOLUTION

36km
FRAME NUMBERS
36km

1
2
3
4

50km STEREO

13
14

64km HIGH RESOLUTION

4-FRAME STEREO SEQUENCE

14-FRAME HIGH RESOLUTION SEQUENCE

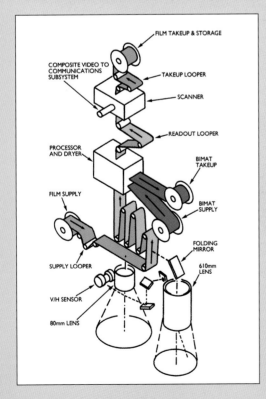

FILM TAKEUP & STORAGE

COMPOSITE VIDEO TO COMMUNICATIONS SUBSYSTEM

TAKEUP LOOPER

SCANNER

READOUT LOOPER

PROCESSOR AND DRYER

BIMAT TAKEUP

FILM SUPPLY

BIMAT SUPPLY

FOLDING MIRROR

SUPPLY LOOPER

610mm LENS

V/H SENSOR

80mm LENS

BELOW Calibration information was pre-exposed alongside the frames of the Lunar Orbiter photographic system's 70mm film. *(NASA)*

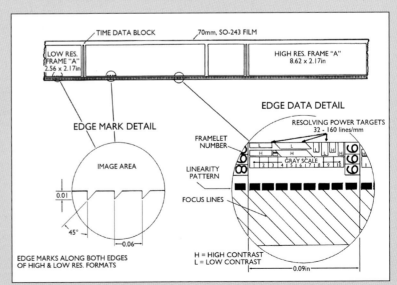

TIME DATA BLOCK

70mm, SO-243 FILM

LOW RES. FRAME "A" 2.56 x 2.17in

HIGH RES. FRAME "A" 8.62 x 2.17in

EDGE DATA DETAIL

EDGE MARK DETAIL

RESOLVING POWER TARGETS 32 - 160 lines/mm

FRAMELET NUMBER

GRAY SCALE

LINEARITY PATTERN

IMAGE AREA

FOCUS LINES

0.01

45°

0.06

EDGE MARKS ALONG BOTH EDGES OF HIGH & LOW RES. FORMATS

H = HIGH CONTRAST
L = LOW CONTRAST

0.09in

off, the used section of the gelatine layer was collected on a take-up reel.

The processed film was dried by running it over a heated drum, then fed to a second looper. It could then either be scanned or pass directly through the scanner to a storage reel to be scanned later whilst winding in reverse. The scanner used a 'flying spot' to produce a signal whose strength represented the density of the emulsion. This data was transmitted to Earth in real time as a frequency modulated analogue signal.

The film was scanned at 287lines/mm. What emerged were rasters, each a 2.54 × 65mm 'framelet' (the calibration data beyond the edge of the image completed the 70mm width) that took 23sec to transmit. The system took 43min to scan a frame pair.

The photography could be terminated at any time by commanding the Bimat to be cut and the film to be scanned as it was being rewound.

It is fascinating to reflect upon the fact that a modern digital imaging system with a solid-state memory, such as carried by the Cassini mission in the Saturnian system, is completely free of such mechanical and chemical complexity and isn't limited by having a fixed length of film. Furthermore, the CCD technology of a digital imager was developed for the spy satellites that superseded those which employed the camera that was adapted for the Lunar Orbiters.

BELOW The scanner of the Lunar Orbiter photographic system produced a succession of narrow 'framelets' spanning the width of the film, including the calibration strip. *(NASA)*

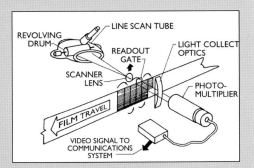

LUNAR ORBITER MISSIONS

Lunar Orbiter 1
Launched on 10 August 1966, it entered a near-equatorial lunar orbit on 14 August. Principal photography was initiated on 22 August and concluded on 30 August. The film scanning and transmission ended on 16 September.

Lunar Orbiter 2
Launched on 6 November 1966, it entered a near-equatorial lunar orbit on 10 November. Photography was initiated on 18 November and concluded on 25 November. The transmitter failed on 6 December, one day before the playback was due to end.

Lunar Orbiter 3
Launched on 5 February 1967, it entered a near-equatorial lunar orbit on 8 February. Photography was initiated on 15 February and curtailed on 23 February, one day earlier than planned due to an intermittent problem with the film transport mechanism – whose drive motor failed on 4 March when 86% through the playback.

Lunar Orbiter 4
Launched on 4 May 1967, it entered a near-polar lunar orbit on 8 May. Photography was initiated on 11 May and curtailed on 25 May due to a problem with the drive. The playback was completed on 1 June.

Lunar Orbiter 5
Launched on 1 August 1967, it entered a near-polar lunar orbit on 5 August. Photography was initiated on 6 August and completed on 19 August. The playback was finished on the 27th.

The feasibility of photographing Earth from lunar distance was discussed a year prior to the Lunar Orbiter 1 mission, but it was not made a mission requirement.

On 22 August, as the spacecraft initiated its principal photography of the eastern end of the Apollo zone, NASA representatives suggested attempting to photograph Earth, even though it would oblige the vehicle to adopt an impromptu attitude. Robert Helberg, the Boeing manager, regarded this as an unnecessary risk, because immediately afterwards the craft would have to re-establish its standard attitude whilst behind the Moon, out of contact with Earth. The company was understandably reluctant, because a large part of the bonus of its 'incentive' contract was reliant on achieving the formal requirements of the primary mission.

Regardless, Lee Scherer, the Lunar Orbiter Program Director at NASA headquarters,

BELOW The geometry of Lunar Orbiter 1's historic picture of Earth on the Moon's limb. *(Lunar Orbiter Image Recovery Project, NASA/Ames)*

Floyd Thompson, the Director of the Langley Research Center which developed the mission, and Clifford Nelson, the Project Manager at Langley, were all in favour.

Boeing relented when the agency agreed to compensate it if the spacecraft were lost as a result of the experiment.

At an altitude of 1,198km, climbing towards the apolune of its almost-equatorial orbit on 23 August, the vehicle was approaching the trailing limb of the Moon as seen from Earth. It turned to point its main engine perpendicular to the line of the limb and then rolled to align the long axis of its narrow-angle camera parallel to the limb. At 16:36:28.6 GMT the shutters fired to obtain a frame-pair.

The narrow-angle frame showed Earth just beyond the lunar limb and an intensely cratered landscape in the foreground.

As our first view of Earth from lunar orbit, this 'tourist shot' was featured in newspapers of the time.

A generation later, progress in the technology of image processing meant that this image could be 'cleaned up'.

The data from the five Lunar Orbiter missions was written on 1,500 magnetic tapes which were stored by the government. When the tapes were reassigned to JPL in 1986, archivist Nancy Evans decided to transfer the data onto a more modern medium.

After a highly specialised Ampex FR-900 tape drive had been acquired and a sample of the raw analogue data read off a tape, it was found that hardware developed for Lunar Orbiter to process the imagery no longer existed so the restoration effort was abandoned.

Interest was rekindled in 2007 when Dennis Wingo, president of the aerospace engineering company SkyCorp, provided initial funds for the Lunar Orbiter Image Recovery Project (LOIRP). This has read all the tapes, digitised the data, and employed modern techniques to reconstitute the images and reveal their true quality.

LOIRP was funded by NASA's Exploration Mission Systems Directorate and Innovative Partnerships Program with support from Odyssey Moon, Skycorp Inc., SpaceRef Interactive Inc., ACES, and the Lunar Science Institute. It was housed at the Ames Research Center in California.

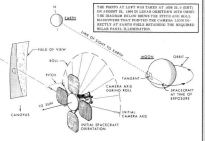

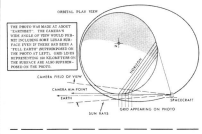

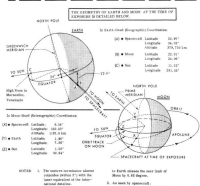

THE PICTURE OF THE CENTURY

A photograph taken by Lunar Orbiter 2 on 24 November 1966 proved the veracity of the saying that 'a picture is worth a thousand words'. It was by far the most striking image of the first two missions.

Interestingly, it was not the kind of target that was originally envisaged.

The film strip had to be advanced at regular intervals to prevent the Bimat from sticking to it. On Lunar Orbiter 1 this had been done with the door of the camera compartment closed. But for the second mission, Tom Young at the Langley Research Center and Ellis Levin at Boeing had devised a procedure for doing it with the door open. A schedule was therefore drawn up to allow this maintenance function to provide useful pictures.

As the mission was being planned, Douglas Lloyd of Bellcomm Inc., which was contracted by NASA to provide mission support, had pointed out that at this particular point there was an opportunity for an oblique view of Copernicus.

The spacecraft was at a height of 46km and the crater, 100km in diameter and 3km deep, was 240km to the north of the ground track.

The illumination was ideal for discerning the nature of the topography – the cluster of central peaks rise 600m above the floor of the crater, the interior of the wall is heavily terraced, and the far horizon is a section of the Carpathian Mountains whose peaks stand 1,000m above the adjacent plain. The clarity of the view is attributable to the lack of an atmosphere. A terrestrial photograph from a similar altitude would be blurred by haze. The character of this terrain isn't readily apparent to a telescopic observer with a vertical perspective.

By providing the vantage point of an astronaut in orbit, it revised our perception of the Moon, revealing it to be a world of dramatic landscapes.

Time magazine quoted NASA scientist Martin Swetnick praising this as 'one of the great pictures of the century'. When the picture was issued to the press, the *New York Times* journalist Walter Sullivan wrote it was 'one of the greatest pictures of the century'. *Life* magazine was definitive, calling it 'The Picture of the Century'.

BELOW The stunning oblique view across the crater Copernicus, taken by Lunar Orbiter 2, which was hailed as the 'Picture of the Century'. *(NASA)*

photographed in an oblique perspective looking west at sunrise to show them as they would appear to astronauts about to land.

The Apollo Site Selection Board reported in March 1967 that the first three Lunar Orbiters, launched in August and November 1966, and in February 1967, had satisfied 'the minimal requirements of the Apollo program for site survey for the first Apollo landing', so NASA released the remaining two spacecraft to the scientists.

Lumpy gravity

Radio tracking of the early Lunar Orbiters in near-equatorial orbits revealed the gravitational field on the near side of the Moon to be irregular. Apart from the variation of velocity with altitude resulting from their being in elliptical orbits, the vehicles kept speeding up and slowing down.

A gravimetric map derived from tracking Lunar Orbiter 5 in a low polar orbit enabled the anomalies to be correlated with surface features. This revealed that the spacecraft was accelerated while it was approaching the 'circular maria' and decelerated immediately afterwards.

As the dark mare plains were low-lying, it was apparent that they must be of a greater density than their surroundings. One early suggestion was that the 'attractor' represented the buried projectile that excavated the basin which was later filled in by mare material but this idea was rejected when it was realised that there were 'negative' anomalies associated with basins which hadn't been filled in. John O'Keefe of NASA argued the attractor was the infill itself. Because the 'positive' anomalies represented concentrations of mass, they became known as *mascons*.

As a depression, a basin might be expected to possess a negative anomaly because a large amount of crustal material had been removed. The sudden removal of crustal material during the excavation of a basin would relieve the pressure on the mantle below and induce deep melting that would in turn cause a plume to rise, lift and fracture the floor, and drive low-viscosity magma to the surface. If enough magma was erupted, its weight would cause the basin floor to sink again. The presence of positive anomalies indicated that when the lava was erupted the crust was too rigid to adjust in an isostatic manner.

A question for later investigation was why most of the basins on the far side of the Moon were still 'dry'. Was the crust thicker in that hemisphere?

The announcement in August 1968 by P. M. Muller and W. L. Sjogren of JPL of lumpiness in the lunar gravitational field was a nasty surprise for the Apollo planners.

RIGHT A map of the equatorial zone of the near side of the Moon showing the various targets examined by the first three Lunar Orbiter missions.
(Harland using NASA data)

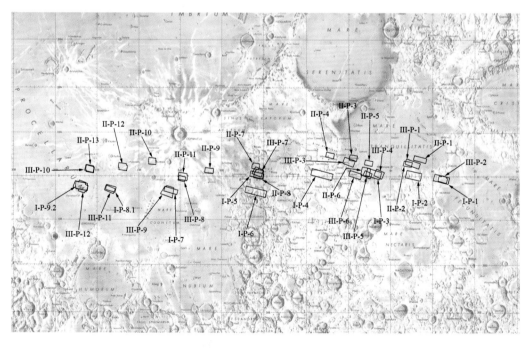

ABOVE Replicas of the Luna 9 spacecraft and its landing capsule. (*Academy of Sciences of the USSR*)

Early Soviet lunar landers

After obtaining the first pictures of the far side of the Moon in 1959, the Soviet Union turned its attention to a spacecraft that could safely deliver instruments to the surface.

It was 1963 before the Lunik series, under the new name of Luna, was ready for testing, and because it was a complex vehicle it suffered a string of failures. The aiming point was in the far western hemisphere to eliminate a horizontal component in the trajectory during the descent.

On 3 February 1966, Luna 9, reputedly the last of the batch, performed perfectly. A rocket fired to slow the spacecraft just above the lunar surface and it ejected a spheroidal capsule which made a 'hard' landing and rolled to a halt. After this had deployed four 'petals' and extended its antennas, its solitary instrument, a TV camera that viewed via a rotating mirror, transmitted a panoramic view of its landing site in the Ocean of Storms.

As luck had it, this transmission was received by the radio telescope at Jodrell Bank in England. A national newspaper fed it into a commercial wire-facsimile machine and produced an image that 'scooped' the official release in *Pravda*. The fresh-looking state of the rocks prompted speculation that the probe had set down on a recent lava flow.

EARLY LUNA LANDING MISSIONS

Luna 4
Launched on 2 April 1963, it was intended to deliver a capsule to the surface of the Moon but failed to make a midcourse correction and the result was an 8,400km flyby on 5 April.

Luna 5
Launched on 9 May 1965, it failed on the way to the Moon and impacted on 12 May at a point in the western hemisphere that is disputed.

Luna 6
Launched on 8 June 1965, it failed on the way to the Moon and made a very distant flyby on 11 June.

Luna 7
Launched on 4 October 1965, it suffered a malfunction in preparing to perform the braking manoeuvre and smashed into the Ocean of Storms on 7 October.

Luna 8
Launched on 3 December 1965, it braked too late and smashed into the Ocean of Storms on 6 December.

Luna 9
Launched on 31 January 1966, it delivered a capsule which transmitted a panoramic view of the landing site in the Ocean of Storms on 3 February at 7.1°N, 64.4°W.

Luna 13
Launched on 21 December 1966, it delivered a capsule to the Ocean of Storms on the 24th at 18.9°N, 62.0°W. In addition to a camera, the capsule had instruments to measure the mechanical and physical properties of the lunar surface.

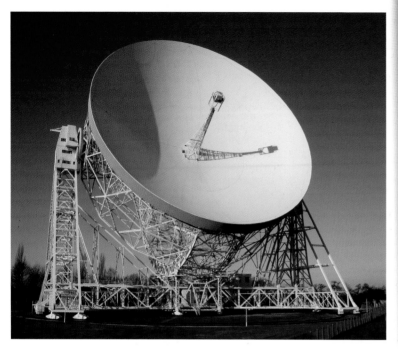

ABOVE The 76m radio telescope at Jodrell Bank. *(University of Manchester)*

ABOVE RIGHT Jodrell Bank eavesdropped on the transmission from Luna 9 and released the first pictures from the lunar surface to a British newspaper. *(Daily Express)*

In America, Harold Urey and Gerard Kuiper were eager for the launch of their own landing mission.

NASA landers

In May 1960, NASA directed JPL to start work on the Surveyor project. Initially it was intended to provide two types of spacecraft, both developed by Hughes Aircraft with a degree of commonality in their systems: one type to enter

lunar orbit; the other to fly an approach trajectory similar to that of the Rangers, then slow itself in order to land. Later the orbiter was reassigned to another part of the space agency and developed without commonality.

Surveyor was considerably more complex than its suicidal predecessor. Its greater weight meant that its introduction had to await the high-performance Centaur stage for the Atlas rocket, the development of which proved problematic.

In view of the many early failures suffered

RIGHT An Atlas-Centaur launches Surveyor 1. *(NASA)*

FAR RIGHT An engineering model of the Surveyor spacecraft in its landed configuration. *(NASA)*

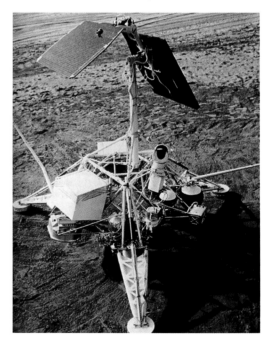

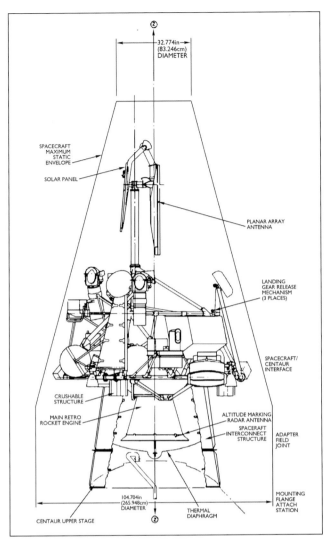

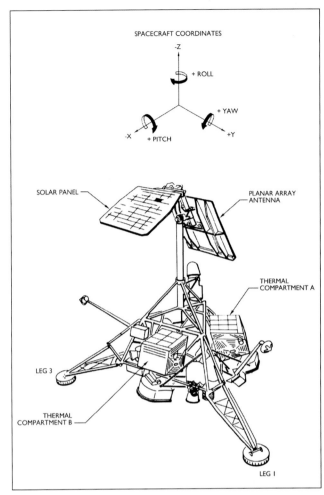

ABOVE Details of the Surveyor spacecraft, including its axes of rotation. *(NASA)*

ABOVE How the Surveyor spacecraft was folded up for carriage in the aerodynamic shroud of its launch vehicle. *(NASA)*

by the Ranger series, when Surveyor 1 was launched on 30 May 1966 it was regarded as simply a test of the basic systems. Its only science instrument was a television camera. The spacecraft deployed its various appendages, made the manoeuvres to aim for its target on the Moon, then ignited its braking system and … touched down safely.

RIGHT Testing the fit of the shroud around the Surveyor spacecraft (left) and the ability of the Centaur stage to command the deployment of the spacecraft's legs and antenna booms. *(NASA)*

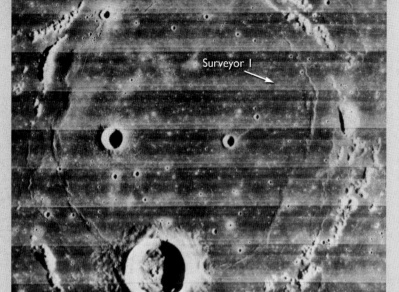

Surveyor 1

IDENTIFYING THE SURVEYOR 1 LANDING SITE

Surveyor 1 took a series of wide-angle frames for a panorama of its landing site.

Raymond Batson of the USGS had conceived a procedure for mosaicking imagery. Prints of the square frames were pasted into their appropriate positions on the interior surface of a sphere. The sphere was then photographed in ten sections, the sections were flat-mounted side by side and then photographed as a single mosaic. In the process, the point of view was altered to cancel the 16° offset of the camera's axis relative to the vertical axis of the lander.

The local horizon was about 2km away. Six features running in a northeast-to-northwest arc were hills beyond the horizon.

Ewen Whitaker of the Lunar and Planetary Laboratory of the University of Arizona identified a number of features on a telescopic picture taken by the university's 155cm NASA-sponsored reflector several months earlier at a time when the illumination was similar. He was able to use triangulation of identifiable features to prove that the landing site was well within the predicted 2-sigma ellipse.

The wide-angle pictures taken by Lunar Orbiter 1 in August 1966 confirmed these terrain identifications and established the position of the lander to within several kilometres.

When Lunar Orbiter 3 imaged the general area on 22 February 1967 the lander was found on one of the narrow-angle frames. It stood out because the shadows cast by crater rims were perpendicular to the Sun line whilst the vehicle cast its shadow down-Sun. At that time, the lander's mast-mounted panels had been oriented to maximise its shadow, so it was observed as a bright object with a thin shadow.

The area was included in Lunar Orbiter 4's mapping mission in May 1967. This enabled the position of Surveyor 1 to be determined in relation to 13 craters whose selenographic coordinates were able to be measured on the global grid, allowing the position to be calculated to an accuracy of plus or minus 0.01°, or roughly 600m. The result was proof that the spacecraft had landed within 15km of the aim point.

Surveyor 1 was the only lander to be documented by a Lunar Orbiter mission.

Gene Shoemaker, head of the camera team, had estimated the chance of a successful landing on the first attempt to be no better than 10%, so he exclaimed, 'My God, it landed!' Engineers at JPL and Hughes had been ready to analyse the downlinked telemetry to attempt to determine why it had crashed.

For this pathfinding mission, the approach to the Moon could be no more than a few degrees off local vertical. The target was near the equator and 45° west of the meridian, inside the Flamsteed Ring, which was a 110km crater that had been 'inundated' by the Ocean of Storms to such an extent that only the taller summits of its rim remained visible.

The mare in the eastern portion of the Ring was darker than was typical for the neighbourhood, and the fact that it appeared to be free of craters, rays, and ridges when viewed telescopically made it interesting to the Apollo selectors.

Given likely uncertainties in the trajectory, the target was a 60km circle in the northern part of the Ring, and subsequent analysis established that the vehicle landed just 14km from the aim point.

The first television image was a wide-angle view of one of the foot pads, which had penetrated a loose material to a depth of only a few centimetres. This was good news for the Apollo engineers, whose heavier lander was to employ much larger pads that would impose the same pressure on the lunar surface.

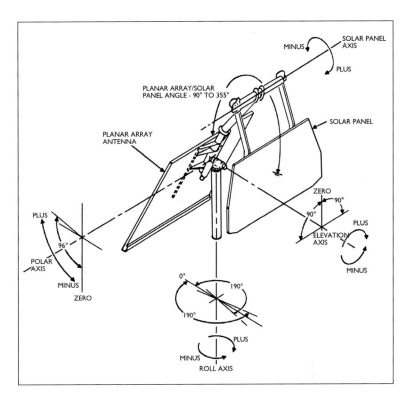

A panorama revealed the lander to be on a gently undulating plain, pockmarked by craters and littered with fragmental debris.

The most prominent crater in the immediate vicinity was 11m distant. It was 3m wide, had a distinct but irregular raised rim, and was about

ABOVE Details of the Surveyor spacecraft's mast. *(NASA)*

BELOW A typical Surveyor trajectory with a direct descent to the lunar surface. *(NASA)*

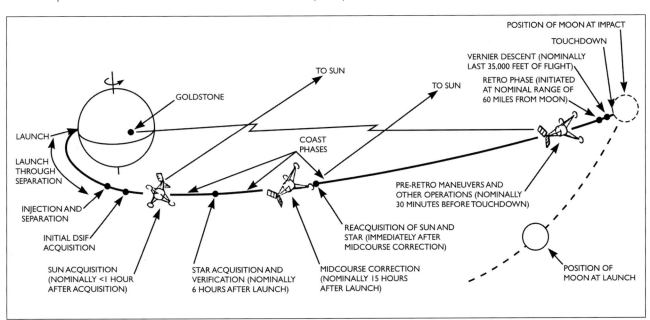

SURVEYOR'S IMAGING SYSTEM

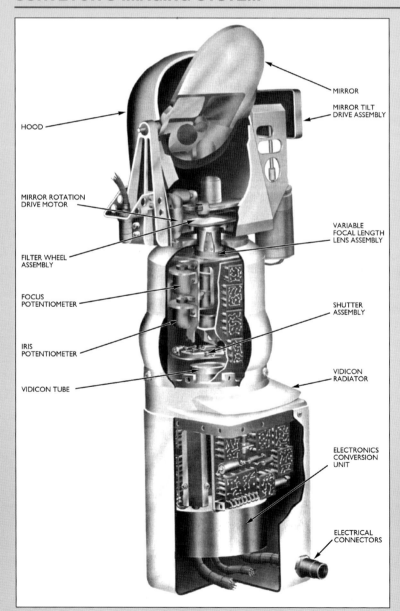

MIRROR

MIRROR TILT
DRIVE ASSEMBLY

HOOD

MIRROR ROTATION
DRIVE MOTOR

VARIABLE
FOCAL LENGTH
LENS ASSEMBLY

FILTER WHEEL
ASSEMBLY

FOCUS
POTENTIOMETER

SHUTTER
ASSEMBLY

IRIS
POTENTIOMETER

VIDICON
RADIATOR

VIDICON TUBE

ELECTRONICS
CONVERSION
UNIT

ELECTRICAL
CONNECTORS

The television camera for the Surveyor lander was mounted on a frame so that it was positioned between two of the three legs.

It consisted of a viewing mirror, filters, lens, shutter, vidicon, and associated electronics. The mirror was an ellipse with axes 15 × 10.5cm that was supported on its minor axis. It had separate drives for motion in azimuth and elevation. The azimuth range was almost 360°, but the elevation range ran from 65° below to 35° above the plane perpendicular to the main axis of the tube. A protective hood rotated in azimuth with the mirror.

The mirror assembly included a wheel with a 'clear' aperture and three filters. A variable-focal-length lens installed between the mirror and the vidicon facilitated focal lengths in the range 25 to 100mm. Although this would normally be operated at its limits for 25.3° (wide) and 6.43° (narrow) fields of view, it could be set for specific focal lengths.

The mirror was located 1.2m above the ground. Since the camera was tilted at an angle of 16° to the lander's mast in order to observe the foot pads of the adjacent legs and the ground between them, when the mirror was tipped at 65° it could view the ground beneath the camera.

There was an iris to allow for differences in illumination of the scene. This offered effective apertures from f/4 to f/22 either in fixed increments that increased the aperture area

LEFT Details of the television camera operated on the lunar surface by Surveyor 1. *(NASA)*

60cm deep. Most of the nearby craters having diameters up to 20m had either low rounded rims or were rimless.

The surface was littered with blocks of rock which had coarse surface textures. Most were angular and were resting on the surface, but some were rounded and appeared to be embedded in the ground. It was inferred that the angular blocks were from a 27m crater 60m away that had a blocky rim. The blocks were single pieces of solid rock that had been excavated from a coherent substrate located beneath a blanket of fragmental material that was at least 1m thick.

In terms of their shapes and distribution, the larger craters in the vicinity were similar to those observed by the Rangers immediately before they crashed. This implied the site was a typical mare surface. It looked to be eminently suitable for an Apollo mission.

Having inspected a western mare site, it was decided to send Surveyor 2 to the Central Bay, a mare plain 170km across that lies on the meridian. Unfortunately, the spacecraft suffered

OPPOSITE A section of the rim of the Flamsteed Ring as observed by Surveyor 1 as hills projecting over its northeastern horizon. *(NASA)*

by a factor of two or in proportion to the average luminance – in the latter case the servo of the iris was controlled by a sensor that used a secondary output from a beam-splitter. The camera settings were to be controlled from Earth but the iris could be set to automatically gauge the illumination of the field of view.

The focal plane was on the vidicon and the image was 11mm square. In the 200-line mode, intended for use with the omni-directional antenna, the vidicon was scanned by an electron beam in 20sec. This required a bandwidth of only 1.2kHz and pictures could be taken at a rate of one every 61.8sec. In the 600-line mode, the vidicon scan and transmission took just 1sec using the high-gain antenna at 200kHz bandwidth. The shooting rate was then only 3.6sec.

In normal use, the mechanical focal-plane shutter was held open for 150millisec and the vidicon would retain the image until it was scanned and cleared. For imaging in Earthlight the shutter could be opened and held while the vidicon was scanned every 3.6sec for 600 lines or 61.8sec for 200 lines. As an option in this mode the vidicon could integrate for a specified duration before the shutter was closed. This was intended to be used to observe objects such as the stars, planets, and solar corona. A sensor would prevent the shutter from opening if the field of view were too bright, but this could be circumvented by transmitting a command from Earth.

The filters were chosen to make the overall camera-and-filter spectral response conform to the standard matching functions of colorimetry. For calibration on the Moon there was a circular photometric chart on one of the foot pads which would be in view of the camera and another on the boom carrying an omni-directional antenna. Each chart had grey-scale wedges around its circumference and colour wedges radially, and there was a perpendicular post in the centre to act as a gnomon whose shadow would indicate the position of the Sun relative to the target.

Upon receipt on Earth, a frame was converted to video format and stored with the appropriate telemetry on magnetic tape. In addition, the video was relayed in real time to JPL, where it was displayed on a closed-circuit television system and archived onto 70mm film.

BELOW Details of the photometric calibration chart carried by the Surveyor 1 lander. *(NASA/Woods)*

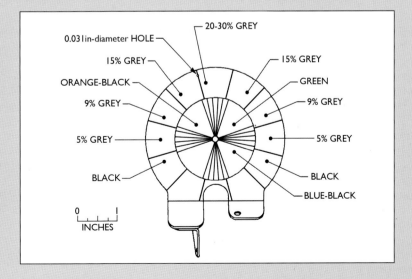

0.031in-diameter HOLE
15% GREY
ORANGE-BLACK
9% GREY
5% GREY
BLACK
20-30% GREY
15% GREY
GREEN
9% GREY
5% GREY
BLACK
BLUE-BLACK
0   1
INCHES

RIGHT A section of a panoramic view taken by Surveyor 1 shortly before sunset which features the spacecraft's shadow. The view spans northeast to southeast, and the outline marked on the horizon marks the hills shown in a previous illustration. *(Courtesy of Philip J. Stooke, International Atlas of Lunar Exploration, 2007)*

BELOW A view by Lunar Orbiter 3 of where Surveyor 3 landed. The box illustrates the area of the enlargement. *(Harland using NASA imagery)*

BELOW An enlarged view of a high resolution frame by Lunar Orbiter 3 showing where Surveyor 3 landed (arrowed, albeit the lander wasn't present when the image was obtained). *(NASA)*

an engine malfunction during the midcourse manoeuvre and impacted out of control near the crater Copernicus.

For the third mission in April 1967 the target was again in the western hemisphere. The 60km target circle in the southeastern Ocean of Storms had been photographed at medium resolution by Lunar Orbiter 1 and at higher resolution by Lunar Orbiter 3 and although there were a number of craters, the mare plain appeared to be fairly smooth.

On a nominal descent, the braking engines would be switched off when the radar reported that the craft was within seconds of contact, but in this case the radar became confused by reflections from rocks and the continuing thrust caused the vehicle to bounce several times before a signal from Earth shut the engines off and it finally settled.

When the television camera started to send back pictures, they revealed that the lander was on a modest slope and that in coming to rest its foot pads had partially embedded themselves in a loose material.

In fact the lander was on the interior slope of a crater that was about 200m in diameter, one of a cluster of such craters. This was welcomed as an opportunity to conduct a detailed study of the interior of a lunar crater by correlating the surface views with the overhead images.

Surveyor 3 was the first to be equipped with a robotic arm which could scrape trenches and lift samples to study the mechanical properties of the lunar surface.

The principal investigator for this experiment was Ronald Scott, an engineer at Caltech, but Hughes had designed and built the hardware.

The results showed that although the lunar material was very finely grained, it was moderately cohesive and, most importantly, its bearing strength increased significantly at a shallow depth; excellent news for the Apollo planners.

Surveyor 4 made a second attempt at the Central Bay in July 1967, but failed. The radio transmission ceased in the final phase of the descent without any indication of an anomaly. The post-mortem identified several things which

BELOW A contour map of the 200m crater in which Surveyor 3 landed. *(USGS/NASA)*

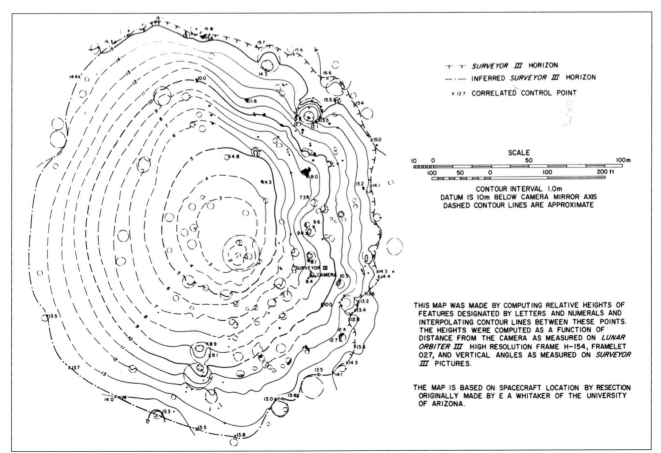

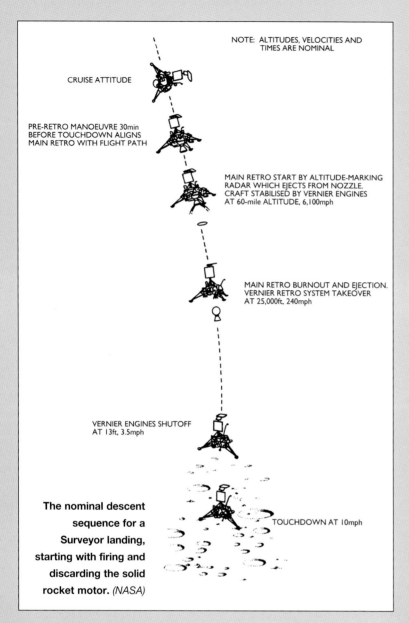

NOTE: ALTITUDES, VELOCITIES AND TIMES ARE NOMINAL

CRUISE ATTITUDE

PRE-RETRO MANOEUVRE 30min BEFORE TOUCHDOWN ALIGNS MAIN RETRO WITH FLIGHT PATH

MAIN RETRO START BY ALTITUDE-MARKING RADAR WHICH EJECTS FROM NOZZLE. CRAFT STABILISED BY VERNIER ENGINES AT 60-mile ALTITUDE, 6,100mph

MAIN RETRO BURNOUT AND EJECTION. VERNIER RETRO SYSTEM TAKEOVER AT 25,000ft, 240mph

VERNIER ENGINES SHUTOFF AT 13ft, 3.5mph

The nominal descent sequence for a Surveyor landing, starting with firing and discarding the solid rocket motor. *(NASA)*

TOUCHDOWN AT 10mph

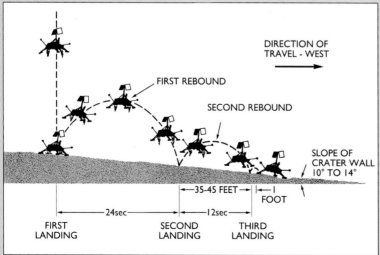

DIRECTION OF TRAVEL - WEST

FIRST REBOUND

SECOND REBOUND

SLOPE OF CRATER WALL 10° TO 14°

35-45 FEET | 1 FOOT

24sec | 12sec

FIRST LANDING | SECOND LANDING | THIRD LANDING

SURVEYOR 3'S BOUNCY LANDING

After discarding the solid rocket motor which had braked its approach velocity, Surveyor 3 fell in the weak lunar gravity until it reached the 'descent contour' that the guidance system was to follow in order to reach the ground by firing its own engines, first to cancel the remaining horizontal component of its velocity and then to control its vertical rate. At an altitude of 300m it was coming down almost vertically with a sink rate of about 30m/sec, and appeared to be home and dry.

But at a height of 10m one of the three angled radar beams lost its lock on the surface. When the flow of data to the closed-loop computer abruptly ceased, the control system reverted to inertial guidance in order to maintain its attitude and throttled the engines to cancel 90% of lunar gravity. Since the radar system was no longer operative, the computer didn't receive the cue at a programmed height of 4.25m to cut off the engines.

The vehicle touched down at a vertical rate of 1.8m/sec with its engines still firing. Although it was level at this time, the surface sloped down to the west and this caused one leg to make contact before the others. In response to this induced tilt, the flight control system, which was attempting to hold a stable attitude and didn't realise it was on the ground, adjusted the throttles to re-establish a level attitude. This increase in thrust lifted the vehicle off the surface.

After peaking at a height of about 11.6m, the vehicle made second contact 15.25m west of the initial point, this time at a vertical rate of 1.2m/sec. As previously, the slope disturbed the attitude and the control system's response resulted in a second lift-off.

When the engines were cut off by a command from Earth, the vehicle had peaked at a height of 3.3m and was descending through 1m.

On contacting the ground for the third time the vertical rate was a mere 0.5m/sec but, as a

LEFT Details of Surveyor 3's tricky landing. *(NASA)*

result of these sideways hops, a horizontal rate of 1m/sec had accumulated. The elasticity of the legs caused the vehicle to rebound and hop another 50cm down the slope before it finally came to rest 11m west of its second point of contact.

The gyroscopes indicated the lander to be tilted towards the west at an angle of 12.5° from vertical, and when the panorama was transmitted it became apparent that the craft was on the inner slope of a crater with a diameter of about 200m.

An investigation concluded that the most likely cause of the radar 'dropping out' was that its logic ordered a 'break lock' as one of its beams crossed a field of rocks. This was because to a microwave radar, angular rocks would appear much as mirror fragments would to a searchlight. The circuitry was designed to ensure the radar tracking would select the strongest signal if several were present. In the final vertical descent, the presence of this scintillating 'side lobe' had therefore obliged the logic to break its lock.

At touchdown, the camera was facing east and the Sun was 11° above the horizon. The fact that many of the early pictures were partially or completely masked by a veiling glare implied that either gaseous engine efflux or fine particles stirred up by the engines during the three 'touch and go' events had coated a portion of the mirror, with the result that the view of the lunar surface was obscured when that section of the mirror was directly illuminated by sunlight. Also, any scene that was strongly reflecting sunlight was similarly degraded.

Later, intermittent sticking of the mirror in both its azimuth and elevation motions indicated that dust had penetrated its mechanism.

These issues impaired the imaging schedule and degraded the results.

The mirror could be rotated in elevation to close the hood and inhibit dust penetration at landing, but the engineers had chosen not to land in that configuration lest the mirror subsequently fail to open and thereby prevent the primary science activities. The hood was improved for later missions.

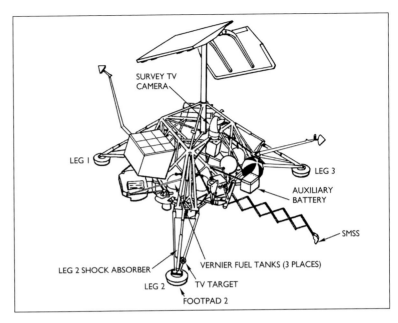

could have gone wrong, but if it was merely a transmitter failure then the spacecraft probably landed safely, albeit mute.

In September 1967 Surveyor 5 was sent to an eastern mare plain. In fact, due to the magnitude of the manoeuvre to cancel the horizontal component of the approach velocity prior to making a vertical descent, this site was about as far east as such a vehicle was capable of venturing. The target was a 60km circle in the southwestern Sea of Tranquillity, southwest of where Ranger 8 had impacted.

A problem during the midcourse manoeuvre

ABOVE The surface sampling arm was installed directly beneath the television camera. *(NASA)*

BELOW A model of the surface sampler. *(NASA)*

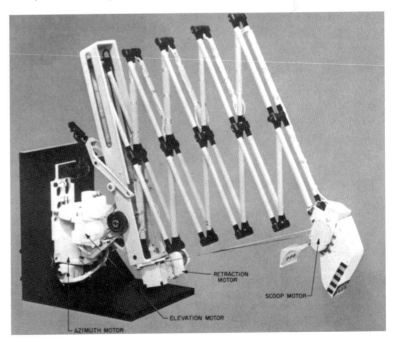

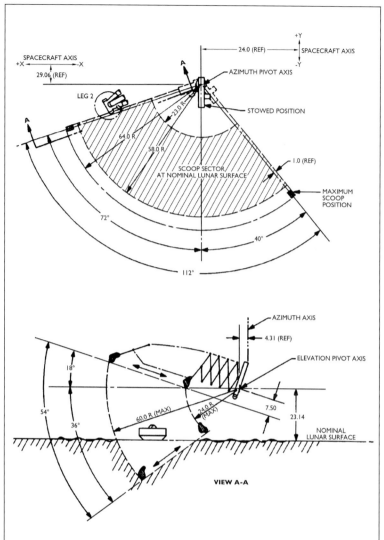

caused a leak of helium pressurant which meant that the engines would lose thrust during the powered descent to the Moon. The vehicle was therefore instructed to begin this descent at a lower altitude, in the hope that it would slow itself sufficiently to make a soft landing. Some engineers estimated the likelihood of success at just 40%. This improvisation was successful, but it was the harshest touchdown to date.

The pictures showed the lander to be on the 20° inner slope of a small and irregularly shaped crater 12m long by 9m wide, from which it could barely see out.

Instead of the mechanical arm, this lander carried an instrument developed by Anthony Turkevich of the University of Chicago to analyse the composition of the surface material. This involved lowering a package which would irradiate the ground with alpha particles (nuclei

BELOW LEFT A television picture of Surveyor 3 operating its surface sampler. *(NASA)*

BELOW Surveyor 3 on the lunar surface, photographed several years later by a visiting Apollo astronaut. *(Harland using NASA imagery)*

RIGHT The camera was modified for the Surveyor 5 mission. *(NASA)*

of helium atoms) and then measure how these were scattered back.

The instrument could measure the abundances of elements that had masses ranging from carbon up to iron, but its capabilities at the heavier end of this range relied on attaining a high signal-to-noise ratio in the energy spectrum.

The data which was transmitted to Earth for a detailed analysis wouldn't be direct evidence of how the elements were combined to create chemical compounds, nor of how these compounds were combined to create minerals; any such insights would be possible only by making assumptions about the nature of the sample.

The three most abundant elements measured were (in decreasing order) oxygen, silicon, and aluminium. This is the same as for Earth's crust.

Due to the unrelenting process of 'gardening' by meteors, the surficial material could be

BELOW The operation of the alpha-scattering instrument's deployment mechanism. *(NASA)*

BELOW RIGHT The operation of the alpha-scattering instrument's head. *(NASA)*

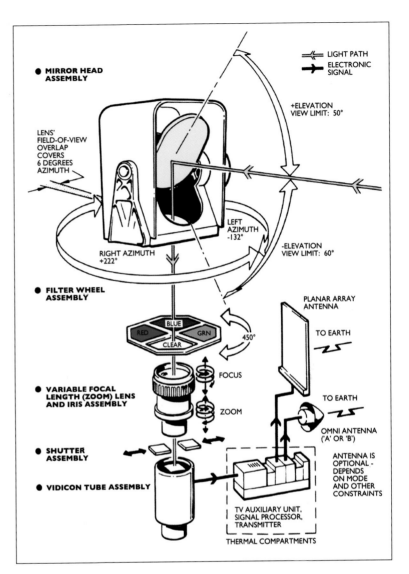

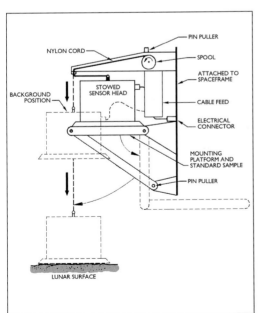

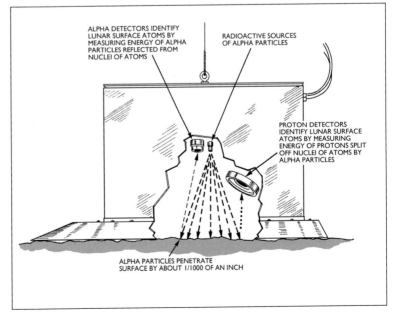

RIGHT A television view of the head of Surveyor 5's alpha-scattering instrument on the lunar surface. *(NASA)*

expected to be primarily fragments of what was beneath (although there would be a contribution from elsewhere) and so it could be presumed to be representative of the bedrock at that locality.

As a chemist, Harold Urey believed the Moon was 'pristine' material condensed directly out of the solar nebula. If this was the case, the 'cold Moon' would possess an *ultrabasic* composition.

The alpha-scattering data indicated the sample was too poor in magnesium to be ultrabasic.

It was also too rich in iron and calcium for it to be an *acidic* rock such as granite.

A good match was a basalt produced by chemical fractionation of ultrabasic silicates.

If it was lava that had been erupted onto the surface from an internal reservoir, then this process of thermal differentiation supported the 'hot Moon' advocated by Gerard Kuiper.

RIGHT A view to the north beyond the crater in which Surveyor 5 landed. *(Courtesy of Philip J. Stooke,* International Atlas of Lunar Exploration, *2007)*

ABOVE Anthony Leonid Turkevich in 1969.
(University of Chicago)

ABOVE RIGHT The surface sampling arm of
Surveyor 7 at work. *(NASA)*

In a paper published a few weeks prior to the Apollo 11 mission, Turkevich boldly predicted on the basis of the Surveyor 5 results that when the astronauts returned with samples from the Sea of Tranquillity, these would prove to be a basalt that was rich in titanium.

Surveyor 6 in November 1967 was the third attempt at the Central Bay, and this time all went well. Like its predecessor, it analysed the composition of the surface material, with broadly similar results.

The sites investigated by the four Surveyors were very similar in terms of topography, and also in terms of the structure of the surficial layer and its mechanical, thermal, and electrical properties.

What struck the scientists as remarkable was the low likelihood that four terrestrial samples which were thousands of kilometres apart and selected in a manner similar to that for choosing the lunar targets, would prove to be so similar.

With this, NASA decided that the project had satisfied its obligation to the Apollo program and, as it had done with the final

SURVEYOR MISSIONS

Surveyor 1
Launched on 30 May 1966, it landed safely on the Moon on 2 June at 2.46°S, 43.23°W and transmitted pictures from within the Flamsteed Ring.

Surveyor 2
Launched on 20 September 1966, it suffered a malfunction on the way to the Moon and then impacted near the crater Copernicus on 23 September.

Surveyor 3
Launched on 17 April 1967, it landed safely in the Ocean of Storms on 20 April at 2.97°S, 23.34°W despite losing its radar lock in the final phase of the descent.

Surveyor 4
Launched on 14 July 1967, its transmission ceased during the final part of its descent to the Central Bay on 17 July.

Surveyor 5
Launched on 8 September 1967, it suffered a propulsion system anomaly during the midcourse manoeuvre but still managed to land safely in the Sea of Tranquillity on 11 September at 1.42°N, 23.2°E.

Surveyor 6
Launched on 7 November 1967, it landed in the Central Bay on 10 November at 0.51°N, 1.39°W.

Surveyor 7
Launched on 7 January 1968, it landed near the crater Tycho on 9 January at 40.86°S, 11.47°W.

RIGHT The northward-looking portion of a panorama taken by Surveyor 7. The outline shows the area in the next illustration. *(Courtesy of Philip J. Stooke,* International Atlas of Lunar Exploration, *2007)*

ABOVE A telephoto view of a succession of ridges to the north of the Surveyor 7 lander. *(Courtesy of Philip J. Stooke,* International Atlas of Lunar Exploration, *2007)*

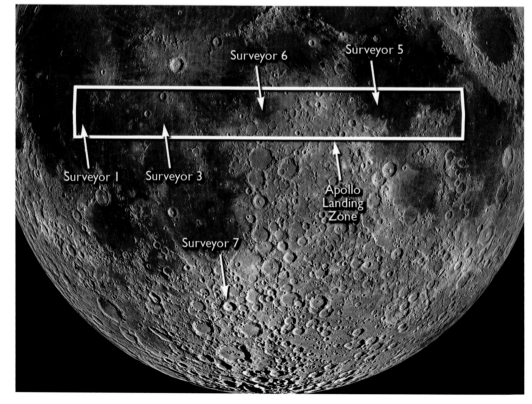

Surveyor 6

Surveyor 5

Surveyor 1

Surveyor 3

Apollo Landing Zone

Surveyor 7

RIGHT The landing sites of the successful Surveyor missions. *(NASA/GSFC/ASU)*

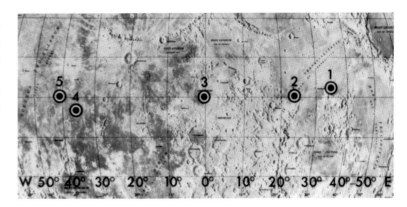

RIGHT Reconnaissance by the Lunar Orbiter missions resulted in five potential targets being chosen for the first Apollo lunar landing. *(NASA)*

CENTRE A map showing the outline of the Lunar Orbiter target designated II-P-6. *(NASA)*

Ranger mission and the two surplus Lunar Orbiters, released the last Surveyor lander to the scientists.

Surveyor 7 was sent far beyond the reach of an Apollo crew, to the northern flank of the 'ray' crater Tycho in the Southern Highlands. It had both the robotic arm and the chemical analyser. This proved fortunate, because when the analyser jammed during its deployment the arm was able to force it down to the ground.

Not unexpectedly, the terrain at this site was much rougher than on the mare plains.

Like sending the final Ranger spacecraft to investigate the crater Alphonsus, this landing in the Tycho ejecta was a tremendous shot in the arm for the science community. Its results will be discussed later.

Overall, despite several losses, the project had achieved its purpose of investigating mare plains in the Apollo zone.

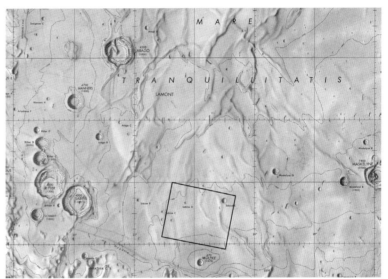

Apollo candidate landing sites

By 1968 a number of candidate sites for the first Apollo landing had been defined as ellipses which spanned 5 × 15km, with the long axis in the direction of approach to cater for uncertainties in the Moon's gravitational field that might cause a vehicle to arrive either 'short' or 'long' of its aim point at the centre of the ellipse.

The eastern option in the Sea of Tranquillity was rejected in favour of the second site, which had the merit of being more typical of the mare plain. The need for three targets spaced 24° apart in longitude placed the backup site in the Central Bay on the meridian and the reserve in the Ocean of Storms farther west.

RIGHT The 15 x 5km ellipse in the II-P-6 area that was selected as the prime target for the first Apollo lunar landing. *(NASA)*

Chapter Eight

Finally, some 'ground truth'

───(●)──────────────

This chapter focuses upon what we learned about the Moon as a result of the two Apollo lunar landings in that momentous year of 1969.

OPPOSITE A boot print in the lunar dust marked the greatest achievement by terrestrial life since it emerged from the ocean. *(Harland using NASA imagery)*

111

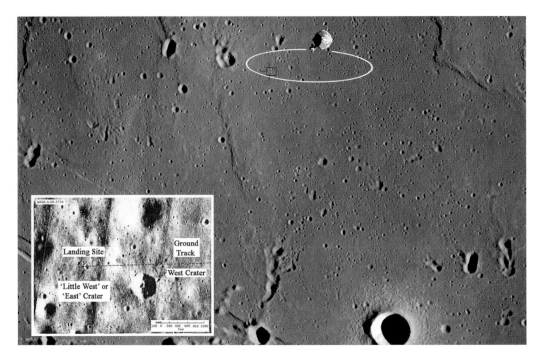

On 16 July 1969, Apollo 11 set off to attempt the first manned lunar landing. The target was a patch of the Sea of Tranquillity whose virtue was that it seemed sufficiently bland to be safe.

The irregular nature of the gravitational field had perturbed the trajectories of the early Apollo missions to enter into lunar orbit. By factoring in the radio tracking from those missions to update the chart of *mascons*, NASA was confident the landing would be on target but the trajectory carried the lander several kilometres west of the intended site.

Lunar dust

Seconds before Neil Armstrong set his craft on the Moon, and just when he really needed to see the surface clearly, it was obscured by dust being blown radially outward by the plume from the engine. On the airless Moon, the dust didn't billow up to envelop the spacecraft, it pursued a shallow ballistic arc for a considerable distance in the one-sixth gravity. His only visual cues were a few rocks that poked up through this thin sheet of dust. So the dust was the first surprise, followed by his amazement at the way it disappeared the moment the engine was stopped, 'just like it had been shut off for a week'.

At the base of the ladder, Armstrong jumped backward and his feet came to rest on the large circular foot pad on that leg of the vehicle.

Noting that the pad was depressed into the surface by several inches, he said, 'The surface appears to be very, very fine grained as you get close to it. It's almost like a powder.'

After transferring his left foot onto the surface, Armstrong made his historic remark, 'That's one small step for a man; one giant leap for mankind.'

Although his boot impressed the surface only a fraction of an inch, largely because most of the dust in this position had been blown away by the engine, there was still sufficient material to make an imprint.

On joining Armstrong outside, one of Buzz Aldrin's tasks was to take a picture of an imprint made by his boot, and he did this west of the lander, where the dust was least disturbed by the engine.

The surface was a fine dust, but unlike the material that Thomas Gold had imagined would flow like a fluid, it was so compacted just below the surface that when Armstrong and Aldrin raised the Stars and Stripes they had difficulty driving the staff into the ground more than 5cm. Similarly, when Aldrin attempted to hammer a hollow tube into the ground to obtain a core sample its penetration fell far short of the intended 40cm.

The dark lunar dust was extremely adhesive, like a fine charcoal. After hauling on a lanyard to transfer items to and fro between the lander

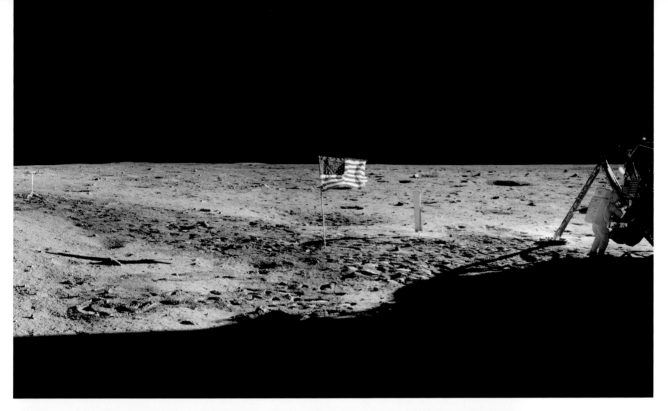

ABOVE A view of Neil Armstrong at the lunar module, showing the flag, the television camera and lots of boot prints. The site was selected for its blandness. *(Harland using NASA imagery)*

LEFT Buzz Aldrin retrieves science equipment from a stowage bay at the rear of the lunar module. *(Harland using NASA imagery)*

RIGHT **When opened in the Lunar Receiving Laboratory, one of the Apollo 11 sample boxes was found to hold 20 rocks with a total mass of 5.5kg.** *(NASA)*

and the surface, the two astronauts' white spacesuits were filthy.

Some scientists had suggested that iron-rich material in the lunar dust might burst into flame on coming into contact with oxygen when the cabin was repressurised, and had expressed concern that the last that would be heard from the crew would be a recital of the checklist leading up to that item!

On lifting their helmets, there was an odour which Armstrong compared to 'wet ashes in a fireplace' and Aldrin to 'spent gunpowder'.

RIGHT **The lunar landing was headline news.**

Rocks

While the moonwalk was underway, Harold Urey, who advocated the 'cold Moon' theory in which the interior was uniformly composed of 'pristine' material from the solar nebula and the maria were splashes of 'impact melt' on a vast scale, was delighted the astronauts didn't report finding the 'frothy vacuum lava' predicted by Gerard Kuiper's 'hot Moon' theory, in which the maria were magma from the mantle which upwelled through fractures.

Unfortunately for Urey, the rocks proved to be a form of basalt that was rich in magnesium and iron. Its texture was strikingly similar to terrestrial basalt. It wasn't impact melt.

This 'ground truth' confirmed that the Moon had undergone thermal differentiation in which lightweight aluminous minerals migrated to the surface and the heavier minerals descended into the interior. The fact that some of this denser material had later erupted onto the surface was incontrovertible evidence that the interior had remained 'hot' for a significant period.

However, in comparison to terrestrial basalts this lunar type was enriched in titanium – just as Anthony Turkevich had predicted based on the Surveyor 5 data. It was also deficient in volatile *alkali* metals such as sodium. This would have endowed its molten form with a low viscosity, enabling it to flow for great distances and create very flat plains. The most striking fact was there was no water chemically bound up in the mineral crystals; the lunar material was *anhydrous*.

Isotopic dating revealed the basalt to have crystallised some 3.84 to 3.57 billion years ago. Chemical differences in the basalt indicated the Sea of Tranquillity had been accumulated by a series of upwellings over an interval of several hundred million years during which the magma reservoir underwent chemical evolution.

OPPOSITE **As Apollo 11 command module pilot Michael Collins said of this picture of the returning ascent stage of the lunar module, it contains the entire human race apart from himself.** *(NASA)*

A sense of timelessness

Afterwards reflecting upon the nature of the lunar surface, Armstrong said, 'My impression was that we were taking a "snapshot" of a steady-state process in which rocks are being worn down on the surface of the Moon with time, and other rocks are being thrown out on top as a result of new events somewhere near or far away. In other words, no matter when you had visited this spot before – one thousand years ago or one hundred years ago, or if you return to it one million years from now – you'd see some different things each time but the scene would generally be the same.'

This was insightful. On the airless Moon there was little chemical erosion. Large impacts simply dug up bedrock which was progressively worn down by smaller impacts to produce the *regolith*, a global blanket of debris consisting largely of very fine dust that had built up over vast ages of time. At Tranquillity Base most of this was pulverised basalt. There was very little meteoritic material in evidence. This process was the 'gardening' that Urey had described after seeing the first Ranger pictures.

Man versus robot

Apollo 11 had unwittingly participated in a race with the Soviet Union to be first to return a lunar sample of Earth.

After Luna 9's pioneering landing, there had been a second such lander and missions where, instead of using its engine to halt its descent, the spacecraft had entered into lunar orbit. On 13 July 1969 the Soviets launched the first mission of their new and considerably larger vehicle.

Luna 9 and Luna 13 had been sent to sites in the far western hemisphere where it was possible to make a vertical landing from a direct approach trajectory. This restriction would not apply to the new vehicle. It was to enter an orbit whose track passed over the target, then restart its engine to descend to the surface.

Luna 15 entered lunar orbit on 17 July, and on 20 July it adjusted its path to create a 16km perilune in the eastern equatorial zone.

Apollo 11 landed on the Moon several hours later. As the moonwalkers rested prior to lifting

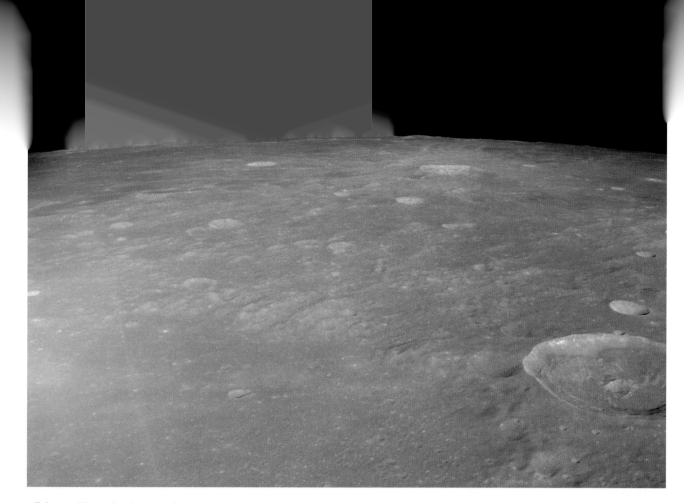

off, Luna 15 crashed attempting to set down in the Sea of Crises.

The Soviet lander's mission had been to land, collect a sample, lift off, and beat Apollo 11 home.

If such a robotic sample return mission could have been accomplished several years earlier it would have been scientifically significant, but in parallel with a human mission which returned with 21kg of lunar material, the resulting small scoop of regolith would merely have provided a useful point of comparison.

The far eastern landing site was necessary because it enabled a craft to lift off on a vertical trajectory and fly directly to Earth; in essence, it was the inverse of landing from a direct descent in the western hemisphere. It was this aspect of the mission profile that required all subsequent Soviet robotic sample return missions to target sites that were too far east for an Apollo crew.

Having sampled an eastern mare site during the first Apollo landing, NASA opted next to investigate a western plain close to the equator, which meant the Ocean of Storms.

The simplest option was the reserve site that had been chosen for Apollo 11, but the

ABOVE The Apollo 12 lunar module in orbit, prior to its descent to the Moon. *(NASA)*

BELOW The objective of Apollo 12 was to land within walking distance of the Surveyor 3 lander. This view looks beyond the inert robot and across the cavity of the 200m crater in which it landed, to the lunar module on the far rim. *(NASA)*

RIGHT Back home, Pete Conrad examines the television camera which was retrieved from Surveyor 3. *(NASA)*

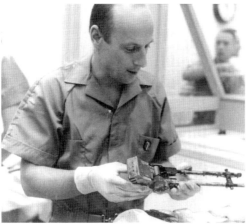

RIGHT Pete Conrad holds the scoop retrieved from the Surveyor 3 surface sampler. *(NASA)*

BELOW A map showing the inferred routes followed by Pete Conrad and Al Bean during their two moonwalks. *(NASA/USGS)*

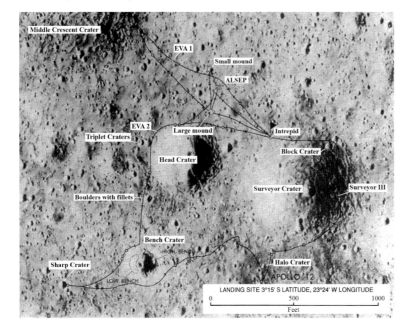

Middle Crescent Crater

EVA 1

Small mound

ALSEP

EVA 2

Triplet Craters

Large mound

Intrepid

Head Crater

Block Crater

Boulders with fillets

Surveyor Crater

Surveyor III

Bench Crater

HIGH BENCH

Sharp Crater

Halo Crater

LOW BENCH

APOLLO 12

LANDING SITE 3°15' S LATITUDE, 23°24' W LONGITUDE

0   500   1000

Feet

APOLLO 12 MISSION

Launched on 14 November 1969, it entered lunar orbit on 18 November. LM 'Intrepid' landed in the Ocean of Storms on 19 November at 3.013°S, 23.419°W. Pete Conrad and Al Bean performed two excursions totalling 7hr 29min, covering a total of 2km, then returned to orbit on 20 November to rendezvous with Dick Gordon aboard the CSM 'Yankee Clipper', after which the ascent stage was crashed by remote control at a speed of 1.67km/sec on a shallow trajectory some 76km east-southeast of the landing site on 20 November at 5.5°S, 23.4°W to provide a seismic signal. The CSM departed on 21 November and returned to Earth with 34kg of lunar material on 24 November.

scientists weren't eager for another bland plain. A scheme to eliminate the trajectory issues that had caused the first mission to land off target now offered the prospect of achieving a 'pin point' landing, so it was decided to send Apollo 12 to set down alongside the crater in which Surveyor 3 resided.

This was achieved by Pete Conrad and Alan Bean, who cut the television camera and scoop off the inert craft to enable engineers to investigate 31 months of exposure to the harsh lunar environment.

During their main traverse, which was a loop covering a total distance of over 1km, Conrad and Bean stopped at several craters with diameters of up to 200m. The crystalline rocks they collected were coarser and more texturally diverse than those from the Sea of Tranquillity. Because they contained less titanium, it appeared in hindsight that the Sea of Tranquillity basalt was unusually *enriched* in this. The variety in basalt chemistry further confirmed that the mare plains couldn't have derived from a single source.

The isotopic ratio measurements yielded an age of 3.2 (±0.2) billion years, indicating 500 million years had elapsed between the eruption of the basalts sampled by Apollo 11 and those by Apollo 12.

Even the most ardent critics of the proposal that the maria formed simultaneously were astonished by such an extended period of volcanic activity.

Analysis of the Apollo 12 samples yielded a major insight into the maria. Paul Gast, head of geosciences at the Manned Spacecraft Center in Houston, found there was potassium, phosphorus, and some of the 'rare earth' elements in the basalts. By linking their chemical symbols, Gast made the label 'KREEP'. Instead of it being present as a new

ABOVE Pete Conrad sampling in the vicinity of Halo crater, using a hand-carried frame to transport their tools and samples. *(Harland using NASA imagery)*

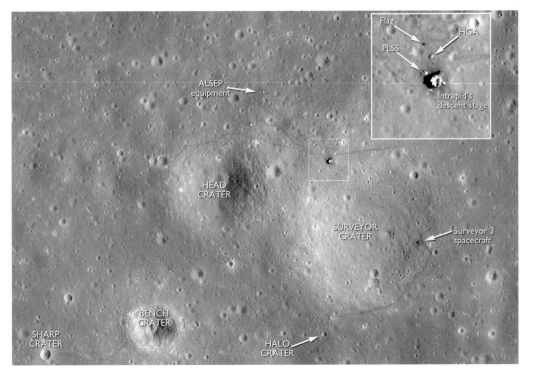

Flag

HGA

PLSS

Intrepid's descent stage

ALSEP equipment

HEAD CRATER

SURVEYOR CRATER

Surveyor 3 spacecraft

SHARP CRATER

BENCH CRATER

HALO CRATER

LEFT A Lunar Reconnaissance Orbiter view of the Apollo 12 landing site showing the discarded descent stage and the routes of the astronauts by how they disturbed the dust. *(NASA/GSFC/ ASU/Woods)*

119

mineral, it was a chemical additive. Hence it is more correct to describe the Ocean of Storms basalts as being KREEPy.

As his 'instant science' explanation at a press conference, Gast suggested this additive might have been 'picked up' from the ancient crust that formed the 'basement' of the mare plain. But when the additive proved to be rich in radioactive elements, in particular thorium and uranium, it was realised that it couldn't be typical of the crust because the heat of radioactive decay would have prevented the crust from solidifying.

The origin of the KREEPy additive therefore became a mystery for a subsequent mission to resolve.

Soviet samples

Although disappointed by the loss of Luna 15, the Soviets bounced back and retrieved their first lunar sample with the next mission in September 1970.

Luna 16 flew a similar trajectory as its predecessor and landed in the northeastern portion of the Sea of Fertility. An arm with a drill rotated down to the surface. Once the drill had taken a 35cm core sample, this was transferred to a capsule in the ascent stage, which promptly lifted off and returned to Soviet territory.

In terms of its titanium content the basalt in the 0.1kg sample of regolith was intermediate between the 'old' Sea of Tranquillity and the 'young' Ocean of Storms, and it was more aluminous. It was dated at 3.4 billion years old.

The Soviets continued this series of missions, but landing a large automated vehicle was a risky venture.

After Luna 18 crashed, Luna 20 managed to set down successfully a few kilometres from the wreckage of its predecessor. Unfortunately, the drill encountered an impenetrable rock at shallow depth and only a small amount of material was able to be recovered.

Luna 23 landed in such a manner that it was unable to operate the new sampler designed to excavate a 1.6m core sample, so the ascent stage was not used.

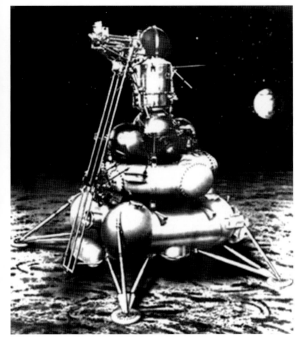

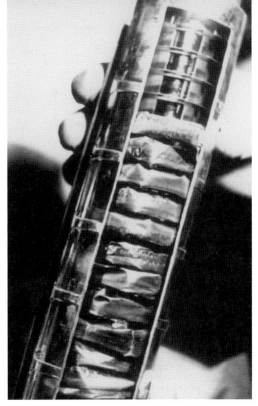

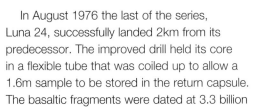

In August 1976 the last of the series, Luna 24, successfully landed 2km from its predecessor. The improved drill held its core in a flexible tube that was coiled up to allow a 1.6m sample to be stored in the return capsule. The basaltic fragments were dated at 3.3 billion years old, implying that the Sea of Crises is one of the freshest of the lunar maria.

With this successful mission, rather ironically achieved not far from where Luna 15 had crashed in its race against Apollo 11, the Soviets concluded their lunar program.

LUNA SAMPLE RETURN MISSIONS

Luna 15
Launched on 13 July 1969, it entered lunar orbit on 17 July and after several adjustments it crashed while attempting to land in the Sea of Crises on 21 July at approximately 17°N, 60°E.

Luna 16
Launched on 12 September 1970, it entered lunar orbit on 17 September and landed safely in the Sea of Fertility on 20 September at 0.68°S, 56.30°E. After it had inserted a 101gm sample in the ascent capsule, that lifted off on 21 September and returned to Earth on 24 September.

Luna 18
Launched on 2 September 1971, it entered lunar orbit on 7 September and crashed attempting to land near the crater Apollonius in the mountains between the Sea of Fertility and the Sea of Crises on 11 September at 3.57°N, 56.5°E.

Luna 20
Launched on 14 February 1972, it entered lunar orbit on 18 February and landed near the wreckage of its predecessor on 21 February at 3.53°N, 56.55°E. Once the 55gm sample had been loaded into the capsule, this lifted off on 22 February and returned to Earth on 25 February.

Luna 23
Launched on 28 October 1974, it entered lunar orbit on 2 November and made a rough landing in the Sea of Crises on 6 November at 13°N, 62°E that precluded operating a drill designed to obtain a 1.6m core sample. As no sample was taken, the return capsule was not launched.

Luna 24
Launched on 9 August 1976, it entered lunar orbit on 14 August and landed within 2km of its predecessor on 18 August at 12.75°N, 62.2°E. The 1.6m drill obtained a 170gm core sample and the capsule lifted off on 19 August, returning to Earth on 22 August.

Chapter Nine

Working on the lunar surface

This chapter focuses on what the later Apollo missions found out about the Moon, how we think the Moon originated, and how the Earth–Moon system is evolving.

OPPOSITE Ed Mitchell has deployed a string of geophones and is walking back along the line, pausing to place a 'thumper' onto the ground to send a seismic shock into the ground. *(Harland using NASA imagery)*

Seismometers on the Moon

NASA began the Ranger project before being directed to send astronauts to the Moon. The idea was that on plunging to its doom, the Ranger spacecraft would release a capsule of science instruments to operate on the surface.

The contract to develop the surface package was awarded to the Aeronutronic Division of the Ford Motor Company in April 1960.

The design mounted the capsule above a braking rocket. At an appropriate altitude a radar altimeter on the spacecraft would command separation. Once it had slowed the capsule's rate of descent sufficiently, the rocket would be discarded.

A spherical 'survival shell' of balsa, a soft material that can readily deform to absorb shock, would protect the 44kg fibreglass capsule when it impacted the ground at 60m/s. As further protection, the 3.6kg scientific payload within the sphere was to be immersed in high-viscosity fluid to absorb the deceleration force. Once the capsule was at rest, the offset centre of mass of the payload would cause it to rotate within the fluid into the orientation for its operation, then the fluid would be drained.

Scientists were eager to obtain insight into the internal state of the Moon and the rate at which meteors strike its surface, so it was decided that the first payload would be a seismometer.

The contract for a battery-powered single-axis seismometer had actually been awarded in July 1959. The experimenters were Frank Press at the Seismological Laboratory at the California Institute of Technology in Pasadena and Maurice Ewing of the Lamont Geological Observatory at Columbia University.

In essence, the instrument was a magnet suspended in a coil by a spring, and restrained radially so that it would respond only to motion parallel to its axis.

Although it would be a considerable technical feat to reach the Moon at all, and in a way it didn't matter where the seismometer was

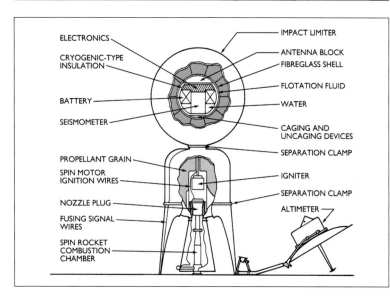

ELECTRONICS
CRYOGENIC-TYPE INSULATION
BATTERY
SEISMOMETER
PROPELLANT GRAIN
SPIN MOTOR IGNITION WIRES
NOZZLE PLUG
FUSING SIGNAL WIRES
SPIN ROCKET COMBUSTION CHAMBER

IMPACT LIMITER
ANTENNA BLOCK
FIBREGLASS SHELL
FLOTATION FLUID
WATER
CAGING AND UNCAGING DEVICES
SEPARATION CLAMP
IGNITER
SEPARATION CLAMP
ALTIMETER

LEFT Detail of the surface package subassembly of the Ranger spacecraft. (NASA/JPL-Caltech/Woods)

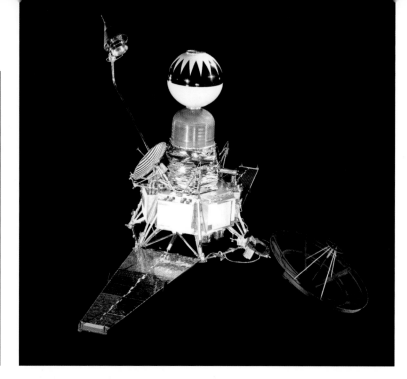

placed, the trajectory was very constrained. The fact that the retro-rocket of the surface package couldn't deal with a lateral velocity component meant the carrier had to perform a vertical descent over the target. This in turn required a site that was close to the equator on the leading hemisphere, in the Ocean of Storms. Furthermore, it couldn't be so far towards the limb that, at an unfavourable libration, the transmission from the surface would be too weak to be read clearly. And of course the Moon had to be visible to the receiver at Goldstone in California when the carrier made its approach.

The Aeronutronic contract called for the first surface package to be delivered by September 1961 for carriage by Ranger 3 in January 1962.

Unfortunately the Atlas rocket didn't respond to steering commands from the ground. It wasn't possible to time the moment of engine shutdown to optimise the final velocity. The 'parking orbit' attained by the Agena upper stage was different from that intended, and the error in the departure trajectory was beyond what the spacecraft itself could correct. It therefore missed the Moon and entered heliocentric orbit.

When Ranger 4 approached the Moon on 26 April it was inert, having malfunctioned during its coast out from Earth. The trajectory provided by the launch vehicle was such that the spacecraft passed around the leading limb and crashed on the far side several minutes later. If it had been functional, the spacecraft could have corrected

its trajectory. NASA hardware had finally reached the Moon, but this was little consolation to the scientists who gained nothing.

The Ranger effort to put a seismometer on the Moon ended with an inert Ranger 5 flying by the trailing limb on 21 October at an altitude of 720km.

But by then NASA headquarters had told the team to delete the surface package and install a television system which could provide the imagery required to reveal the nature of the

ABOVE A replica of Ranger 3 with the surface package subassembly mounted on the main hexagonal frame, the high gain antenna deployed, and the boom of the low gain antenna rotated off the apex of the surface package.
(NASA/JPL-Caltech)

LEFT Preparing the surface package subassembly of a Ranger spacecraft.
(NASA/JPL-Caltech)

ABOVE **The original Space Flight Operations Center at JPL during the Ranger 5 mission.** *(NASA/JPL-Caltech)*

BELOW **Buzz Aldrin stands beside the seismometer that he has just deployed.** *(Harland using NASA imagery)*

lunar surface at close range. In essence, the project had been commandeered to support the Apollo program.

But all was not lost, because the space agency also ordered that a package of scientific instruments be developed for Apollo astronauts to deploy on the lunar surface. The package needed to be capable of being offloaded from the lander and set up by two astronauts within an hour, and was to transmit data to Earth for at least one year.

In 1964 several science planning teams drew up preliminary lists of instruments. Development funding was allocated in June 1965. The project became the Apollo Lunar Surface Experiments Package (ALSEP). In March 1966 the contract to design, manufacture, test, and supply a series of ALSEPs went to Bendix of Ann Arbor, Michigan.

A modular concept would allow packages for particular missions to be selected on a 'pick and mix' basis.

In late 1968 it was decided that because the first Apollo lunar landing would be essentially an engineering test flight with extremely

limited time out on the surface, it wouldn't be assigned an ALSEP.

When the scientists complained, the agency said the first mission could deploy a simpler package. With a seismometer still having the highest priority, it was decided to modify the ALSEP instrument to draw its power from a pair of solar panels rather than from the central generator unit made for a fully integrated package.

At 48kg this seismometer was heavier than the entire Ranger surface package, and rather more capable than that previous instrument because in addition to a single-axis short-period sensor for vertical motions at frequencies with a resonant period of about 1sec, it had a trio of orthogonal sensors with longer periods that would measure surface motions in both the horizontal and vertical directions.

Deployed by Buzz Aldrin, the seismometer operated through the remainder of the lunar day, surviving the heat, and was commanded to shut down just before sunset. It was restarted after sunrise but the electronics had been damaged by the intense cold during the long lunar night and the transmission was poor. Near noon of its second day the instrument failed. Nevertheless, the experiment had established that seismometry could be performed on the lunar surface.

However, it was not until Apollo 12 that scientists were able to really interpret the intriguing results from the first instrument. On this mission, the spent ascent stage of the lunar lander was deliberately crashed on the Moon. The ALSEP seismometer reported the crust 'ringing' for almost an hour with a signature that was unlike any terrestrial signal.

Gary Latham, the principal investigator at the Lamont Geological Observatory at Columbia University, noted that only with this 'calibration' did it become possible to distinguish between a moonquake and an impact. Surprisingly few seismic events were of internal origin.

To provide stronger calibration signals, it was decided that on future missions the spent third stage of the Saturn V launch vehicle would be manoeuvred to strike the Moon far more energetically than could be achieved by crashing an ascent stage.

Between them, the Apollo 12, 14, 15, and 16 missions assembled a network of seismometers to triangulate the locations of meteoroid strikes

and the epicentres of moonquakes. The results established that the crust was extremely fragmented to a depth of about 35km, forming a *megaregolith*.

In comparison to Earth, the Moon is almost seismically inert; barely one billionth of its activity. Most of the events that were detected were only magnitude 1 or 2 on the Richter scale; on Earth these would be ignored as merely the 'noise' of the continuously adjusting crust. The largest event ever detected on the Moon before the network was switched off in 1977 was magnitude 4.

The largest impactor was estimated to have had a mass of 5 tonnes. This landed on the far side and the propagation of the seismic waves served to probe the interior structure of the globe. The analysis confirmed the Moon to possess a crust, a mantle, and a core. This was only to be expected of a thermally differentiated body, just as Gerard Kuiper had claimed and Harold Urey had denied.

Some internal seismic events were near the surface, but most originated at depths of 800 to 1,000km and there was a strong correlation

| Apollo 13 S-IVB | Apollo 14 S-IVB | Apollo 15 S-IVB | Apollo 17 S-IVB |

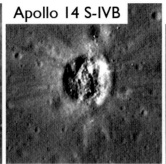 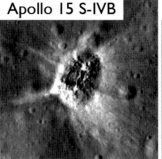 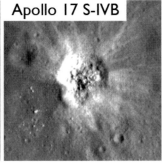

ABOVE When the impact craters made by spent Apollo S-IVB stages were identified by Lunar Reconnaissance Orbiter they all proved to be very similar, at about 35m wide. *(NASA/GSFC/Arizona State University)*

with the tidal forces that are imposed on the Moon by its elliptical orbit of Earth.

The damping out of seismic energy below a depth of 1,000km indicated this zone was semi-molten, being 'warm' rather than 'hot'. Although this layer cannot be deep enough for convection currents to occur, it might contain isolated pockets of fluid and the 'deep' seismic events might be due to magma moving in response to tidal forces.

Magma ocean

Whilst most of the material in the regolith of the Sea of Tranquillity returned by Apollo 11 for analysis was a pulverised basalt, it contained a small residue with a very different character.

On the basis of the chemical analysis performed by Surveyor 7 close to the crater Tycho in the Southern Highlands in 1968, Gene Shoemaker had predicted that 4% of the regolith sampled by the first Apollo crew would comprise minuscule fragments of light-coloured rock. This proved to be the case. The principal mineral was *plagioclase feldspar*.

Terrestrial plagioclase is rich in sodium but the Moon is depleted in sodium and the lunar variant had calcium, making it *calcic-plagioclase*. Some of the fragments were sufficiently pure to justify their being called *anorthosite* (the term for a rock that is at least 90% plagioclase) but most were diluted with iron-rich minerals and therefore were more properly referred to as *anorthositic gabbro*, which was what Surveyor 7 had found.

Anorthositic rocks were widely scattered across the Moon in the form of bright 'rays' from impacts in the highlands. And since it was now evident that the maria were created when magma from the mantle erupted from the

fractured floors of impact basins, the highland terrain was representative of the original crust. Now scientists had small samples of this material to study in their laboratories.

At a conference in early 1970 to announce the results of analysing the Apollo 11 samples, John Wood of the Smithsonian Astrophysical Observatory reported that the density of the anorthositic material, at 2.9g/cc, versus 3.4 for the Moon as a whole, meant that the heat released by the giant impacts during the accretion of the Moon from planetesimals must have created a global *magma ocean* which later crystallised to produce the primordial crust. This was a major insight into early lunar history.

Sampling the Imbrium basin rim

After Apollo 12 achieved a 'pin point' landing, NASA reduced the target ellipse to enable the next mission to aim for a more confined site in rougher terrain. To escape the confinement of the equatorial zone, it also had to relax the propellant margins. And the requirement to define a backup site was cancelled. A launch delay of up to three days would be accommodated by landing at the higher Sun angle. This was a welcome degree of flexibility. Shrinking the ellipse and eliminating the rule that the general area needed to be free of terrain relief permitted more scientifically interesting sites to be considered.

It was decided to send the next mission to the hummocky terrain that forms much of the periphery of the Imbrium basin. This Fra Mauro Formation is actually the most extensive stratigraphic unit on the near side of the Moon, although it is not contiguous because in places it is interrupted by patches of mare, craters, and their blankets of ejecta.

Geologists had devised a timescale for lunar history in terms of how the ejecta from the Imbrium impact splattered across thousands of kilometres, in the process sculpting ruts and grooves. But this was a *relative* scale because they didn't know when this momentous event occurred. Dating this impact was therefore the most important item on their agenda.

During the 'hinge' process by which an impact produces a crater, the material from the deepest point of excavation is dumped right on the rim and the shallower material is distributed progressively farther out. To sample the blanket of Imbrium ejecta lying beneath the 'gardened' surface of the Fra Mauro, the geologists sought a crater that would serve as a natural 'drill hole'.

This had to offer a clear line of approach from the east, a landing point within a kilometre or so, be recent, and possess a very blocky rim for sampling. They chose a 370m wide pit atop the crest of a north-south ridge some 40km north of the crater which had given its name to the overall formation. Given its steep interior walls, the sampling target was named Cone Crater.

Because this was precisely the kind of terrain that had been avoided in searching for 'safe' sites for the earlier landings, NASA had only a few high resolution pictures which had been taken by Lunar Orbiter 3 for scientific interest. After Apollo 12 obtained pictures, it was confirmed on 10 December 1969 as the target for Apollo 13.

The best place for a landing was the relatively flat plain 1km west of the ridge but this was deemed to be uncomfortably close to Cone's ejecta, so a position slightly farther west was selected as the computer's aim point.

Unfortunately, Apollo 13 had to abort on the way to the Moon because of an explosion.

The site was deemed so important that it was assigned to Apollo 14 and, after overcoming some technical gremlins, Al Shepard and Edgar Mitchell set down within 50m of the aim point, making theirs the most accurate landing of the entire program.

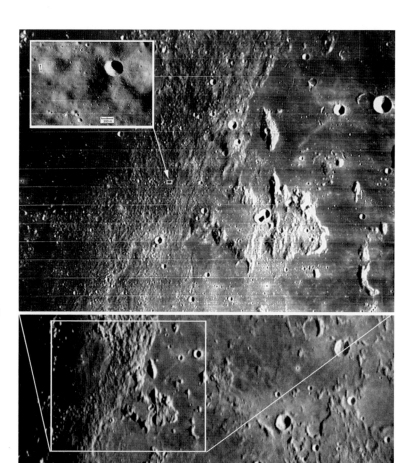

ABOVE The target for Apollo 13 was the hummocky formation to the north of Fra Mauro, the large and eroded crater near the terminator of the telescopic photograph (bottom). The area had been viewed by Lunar Orbiter 3 as a secondary target of scientific interest (main image). The inset shows the location of Cone Crater. *(Harland using NASA imagery)*

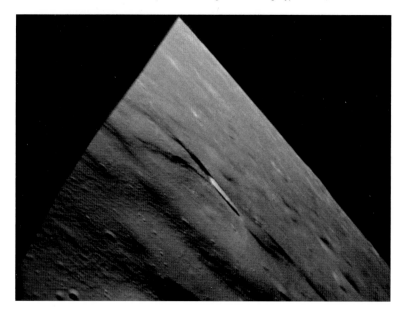

RIGHT A still image taken from a 16mm movie shot from the Apollo 14 lunar module as it approached the ridge on which Cone Crater was located. The aim point for the landing was on the plain about 1km beyond. *(NASA)*

APOLLO 13 MISSION

Launched on 11 April 1970, it had to abort after 55hr and return to Earth. The S-IVB stage was crashed on a steep trajectory at a speed of 2.83km/sec on 15 April at 2.4°S, 27.86°W to provide a calibration signal for the seismometer installed by Apollo 12 some 137km to the east-southeast.

APOLLO 14 MISSION

Launched on 31 January 1971, it entered lunar orbit on 4 February. LM 'Antares' landed at the target site in the Fra Mauro Formation on 5 February at 3.645°S, 17.471°W. Al Shepard and Ed Mitchell performed two excursions totalling 9hr 23min and walked a total distance of 3.3km. They returned to orbit on 6 February to rendezvous with Stu Roosa aboard the CSM 'Kitty Hawk', after which the LM ascent stage was crashed on a shallow trajectory at 3.42°S, 19.67°W, between the Apollo 12 and Apollo 14 landing sites, on 7 February to provide a seismic signal. The CSM departed on 7 February and returned to Earth with 43kg of lunar material on 9 February. The S-IVB stage was crashed on 4 February on a steep trajectory at 8.1°S, 26.6°W to provide a seismic signal.

BELOW A map showing the inferred routes followed by Al Shepard and Ed Mitchell during their two moonwalks. The inset looks beyond the lunar module to the ridge that was their objective.
(NASA/USGS)

For Apollo 14, the ALSEP included an active seismic experiment. Mitchell laid a cable that incorporated a number of geophones. Then he walked back along the line, pausing at 3m intervals to set a 'thumper' on the ground and detonate a small charge (essentially a shotgun cartridge) to send a seismic signal into the ground. The data picked up by the geophones was radioed to Earth.

Although 8 of the 21 charges failed to detonate, the experiment was successful. The data showed that the regolith at the landing site was 8.5m deep and that the local Fra Mauro Formation blanket was of the order of 75m thick. It was evident that with Cone on the crest of the 100m tall ridge, the crater had almost certainly not penetrated into whatever was beneath this blanket, confirming that the rocks sampled on its rim would indeed be Imbrium ejecta.

On their second day, Shepard and Mitchell set off east towing a two-wheeled cart of apparatus.

In contrast to the level plains explored by their predecessors, the surface was undulatory with a vertical amplitude of 2m. It was all too easy to become disorientated.

The scientists had marked a number of sites along a planned route at which the astronauts were to take measurements and collect samples, but the craters that were the principal navigational features on the map were invisible at ground level and a lot of time was wasted in trying to find these specific targets.

Pushing on, they started up the ridge, whose slope soon increased to 10° and they began to huff and puff. Upon reaching the crest with time running out, they were unsure of their location.

They had hoped to identify Cone Crater by its raised rim but couldn't see such a feature. However, there was a large field of boulders and they decided to sample those in the time remaining. In doing so they made their way towards the largest boulder, and after sampling this they headed home.

A post-flight study established that this field of boulders extended right to the lip of the crater, and that the astronauts had ventured within 15m of the lip. Nevertheless, by returning with a football-sized rock which was precisely what they had travelled so far to retrieve, they had achieved their primary objective.

Analysing the Apollo 14 rocks was rather

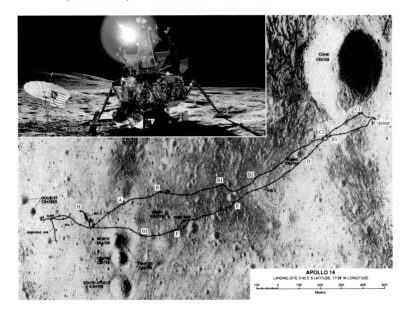

RIGHT On the way to Cone Crater, Al Shepard obtained a core sample alongside the two-wheeled modular equipment transporter. *(Harland using NASA imagery)*

CENTRE The large boulder that was sampled close to Cone Crater, the rim of which is just 15m beyond. *(Harland using NASA imagery)*

more complicated than was the case for those from the maria, because the rocks from Fra Mauro were *breccias*, comprising shattered precursor rocks consolidated within a matrix of finely grained material. The simplest breccia was composed of fragments of a single precursor, but if breccias were smashed and then reformed in new configurations, the inclusions in a breccia, known as *clasts*, might themselves be complex structures.

Dating when the breccias were bound together in order to date the impact which applied the shock was complicated by the fact that when a rock is melted, this 'resets' the isotopic 'clocks' used in determining the formation date. This wasn't an issue for basalts, but analysing breccias involved dating the individual clasts.

The samples tended to cluster into two ranges, one 3.96 to 3.87 billion years and the other 3.85 to 3.82 billion years. It was presumed the older dates, derived from some of the clasts, were the formation ages of the rocks which were later shattered to make the breccias. From this it was inferred that the Fra Mauro Formation was a splash of very hot Imbrium ejecta that was deposited between 3.85 and 3.82 billion years ago.

A study of the Fra Mauro impact-melt breccias revealed they were KREEPy, being rich in both radioactive and 'rare earth' elements. This chemical additive had crystallised during the final phase of the fractionation of the material in the magma ocean. Let's consider that concept for a moment.

RIGHT In the Lunar Receiving Laboratory Ed Mitchell (left) and Al Shepard examine the 3.4kg 'football-sized' breccia that they collected from the field of boulders near Cone Crater. *(NASA)*

ABOVE An Apollo 15 view from orbit of Hadley Rille and the craters Autolycus (near) and Aristillus (beyond) on the Sea of Rains. The first probe to hit the Moon, Lunik 2, impacted somewhere in the vicinity of Autolycus. *(NASA)*

In the process of crystallisation from a melt, certain elements are either accepted or rejected according to whether they are able to fit into the regular crystalline lattice. Because trace elements tend not to participate in mineralisation, they stay in the melt while the compatible elements are extracted to become part of the solidified rock. As a result, the incompatible elements become more concentrated as they remain in the diminishing pocket of molten rock. The radioactive elements at depth would have helped to maintain this concentrated reservoir in a molten state, and then been locked in when it finally solidified. Some process had later brought this material to the surface in the Moon's western hemisphere.

NASA decided in September 1970 that Apollo 15 would be the first mission to use a Lunar Roving Vehicle (LRV), which enabled a 'multiple-feature' site to be assigned for exploration during three wide-ranging excursions.

After realising that the primordial crust of the Moon was anorthositic in composition, the main scientific objective for Apollo 15 was to obtain a sample of this material *in situ*.

It was decided to seek this on the flank of a mountain in the Apennine range that traces the southeastern perimeter of the Imbrium basin.

A number of landing sites were considered, but the selectors chose a bay of the Sea of Rains, within the basin, in order to be able to sample another mare site.

As a bonus, the astronauts would be able to visit Hadley Rille. Being one of the Moon's most impressive sinuous canyons, this remarkable feature originates in an arcuate cleft in the basin-facing side of the Apennines. It exploits a network of radial and peripheral clefts to follow the mare shore north for 10km until, near the base of the 3,500m Mount Hadley Delta, it cuts across the mouth of the bay then resumes its sinuous path to end on the mare plain. At the selected landing site, the rille was more than 1km wide and 350m deep, with steep walls and a flat floor that was littered with enormous boulders.

The approach was tricky because it required the lunar module to fly over the mountains before initiating a steep descent to land east of the rille. But thanks to the mobility offered by the LRV, it wasn't essential to set down at a specific point.

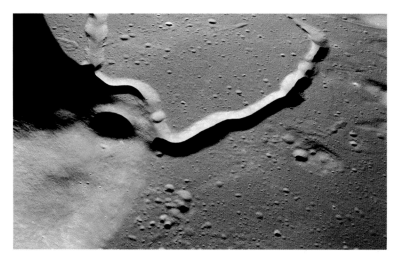

LEFT A view of the Apollo 15 landing site taken by the spacecraft prior to making its descent. Mount Hadley Delta is on the left and the aim point was just short of the rille. *(Harland using NASA imagery)*

On driving to the rim of the rille, Dave Scott and Jim Irwin saw thick horizontal layers of lava exposed in the opposite wall. It was the only occasion during the Apollo program that astronauts were able to observe how the lava flows of the maria were stacked one atop the other.

On their second excursion, Scott and Irwin drove south several kilometres to Mount Hadley Delta.

The fact that the lunar mountains were well rounded with smooth flanks instead of being the

ABOVE Dave Scott attends the LRV with Hadley Rille as a backdrop. The canyon is over 1km wide and 350m deep. *(Harland using NASA imagery)*

APOLLO 15 MISSION

Launched on 26 July 1971, it entered lunar orbit on 29 July. LM 'Falcon' landed at Hadley-Apennine on 30 July at 26.132°N, 3.634°E. Dave Scott and Jim Irwin performed three excursions totalling 18hr 33min, driving 27.9km, then returned to orbit on 2 August to rendezvous with Al Worden aboard the CSM 'Endeavour', after which the LM ascent stage was crashed on a shallow trajectory 93km west of the landing site on 3 August at 26.36°N, 0.25°E to provide a seismic signal. The CSM departed on 4 August and returned to Earth with 77kg of lunar material on 7 August. The S-IVB stage was crashed on 29 July on a steep trajectory at 1°S, 11.9°W to provide a seismic signal.

BELOW A telephoto view across Hadley Rille. The near rim is marked by the boulders in the foreground. There are intermittent lines of outcrops high on the far wall and scree littering the slope below. *(Harland using NASA imagery)*

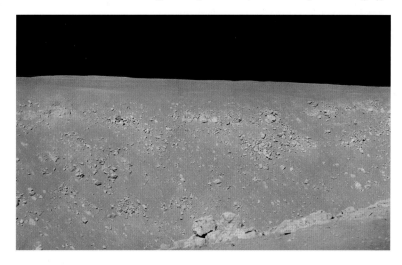

sharp peaks imagined by science fiction authors and movie-makers, was because loose material flowed downslope. As a result, the transition from the plain to the flank of the mountain was gradual and the LRV was readily able to climb to an elevation of several hundred metres.

Unlike *orogeny* (mountain building) on Earth, where the process of plate tectonics causes continents to collide and produce mountain ranges on a timescale of millions of years, the mountain chains of lunar basins were created in an instant when the shock of the impact broke the crust into large blocks which were displaced and tilted.

By sampling the rim of a 'drill hole' crater on the flank of the mountain, it was hoped to find a piece of the underlying crustal block. This was to be achieved by locating a crater on the lower flank that was sufficiently large to have dug through the loose material that had slipped downslope.

The largest crater in the assigned sampling area was named Spur, and it had a diameter of about 100m. In walking along its rim, Scott and Irwin discovered a white rock that was perched atop a small pedestal of darker material. Upon lifting the white rock and examining it, they saw crystals sparkling in the sunlight and knew they had found what they had been sent to find.

The task of Apollo 15 had been portrayed by reporters as being to find the oldest rock on the Moon, and when a scientist at a press

conference sought to explain how this particular rock would advance the study of the origin of the lunar crust and referred to petrogenesis, the journalists promptly labelled it the Genesis Rock.

Analysis confirmed that, at 98% plagioclase, the 0.27kg rock was indeed anorthosite. Aged at 4.1 (±0.1) billion years, it certainly predated the creation of the Imbrium basin some 3.85 to 3.82 billion years ago. Its crystalline structure wasn't in pristine condition because it had suffered intense shock, but the material was clearly derived from the aluminous crust of the magma ocean.

While on the flank of Mount Hadley Delta the astronauts encountered a large breccia boulder that had acquired a coating of greenish material on one side. Laboratory tests showed that this coating consisted of microscopic droplets of glass produced by a 'fire fountain'.

Such an eruption was likely to occur as the precursor to an extrusion from a deep fracture in the floor of an impact basin. As magma from the mantle reservoir rose through the crust it would liberate the volatiles which were in solution, and these would blast out under pressure as a mist of small droplets. If magma is allowed to cool slowly, it crystallises but if it cools rapidly it creates glass. In the vacuum of space, the mist issued from a fire fountain would shock-cool to create glass spherules. On the airless and low-gravity Moon such a plume would rise to a height of several hundred kilometres and then fall back to make a pyroclastic blanket spanning a wide area.

The glass in this case was green owing to the magnesium-rich silicates it contained. The material would have risen through the crust so rapidly that it wouldn't have had time to evolve chemically. It was therefore a welcome 'pristine' sample of the olivine and pyroxene of the mantle. Dating established that the fire fountain occurred 3.3 billion years ago.

Seeking recent volcanism

Scientists have long debated whether the Moon is still volcanically active.

Amateur astronomers have reported seeing anomalous 'glows' and patches of 'obscuration' known as *transient lunar phenomena*, but such observations are subjective.

After reports of a 'veiling' of the floor of the large

LEFT **In the Lunar Receiving Laboratory, Dave Scott examines the 'Genesis Rock'.** *(NASA)*

BELOW **The 'green boulder' on the flank of Mount Hadley Delta, with a view across the plain to Mount Hadley (now fully illuminated) whose summit is 24km away, with other Apennine peaks beyond.** *(Harland using NASA imagery)*

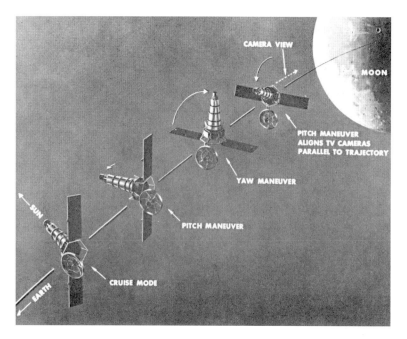

mare plains, NASA released the final spacecraft of the series to the scientists, who sent it diving into Alphonsus in search of volcanoes.

As Ranger 9 made its approach on 24 March 1965 it performed a final manoeuvre to refine its aim, then faced its battery of cameras along the velocity vector to maximise the resolution of the final frames.

As an innovation, the Ranger video downlink was converted to a format suitable for live broadcast by the American TV networks. This was done by using two pairs of vidicon tubes, one for the wide-angle stream and the other for the narrow-angle stream. In each case, one vidicon viewed the image that was displayed on its counterpart, in the process converting the 1,132 lines per frame being received from the spacecraft into the 500 lines of the commercial system. The result impressed not only the public but also scientists who hadn't previously appreciated the value of imagery.

The transmission was begun at an altitude of 2,400km, lasted 18 minutes, and ended with the spacecraft impacting at 9,617km/hr within 6km of the aim point. Surface detail as small as 25cm could be resolved when the final frame was analysed.

Contrary to expectation, there was no clear evidence that the central peak of Alphonsus was a volcano. Gerard Kuiper argued that small craters on the floor that possessed dark 'halos' were volcanic in nature but still the evidence was subjective.

In 1966 Barbara Middlehurst of the University of Arizona published a study of transient phenomena and pointed out that their frequency tended to increase when the Moon was near perigee and apogee, hinting at a correlation with the gravitational stresses imposed on the Moon by Earth. On the other hand, the fact that there were many hundreds of reports of 'brightenings' and 'obscurations' was argued by sceptics to be evidence that they were illusory.

crater Alphonsus by Dinsmore Alter in the USA in 1956, Nikolai Kozyrev in the USSR kept occasional watch using the 1.2m reflector of the Crimean Observatory and on 3 November 1958 he obtained a spectrogram which he said indicated the presence of a cloud of glowing molecular carbon, but the detection was disputed.

After the Ranger project had inspected two

LEFT A selection of the central portions of nested frames running left to right and top to bottom from the A-camera of Ranger 9. The first frame here was taken at an altitude of 1,600km and the final frame at 7.2km. (Harland using NASA/JPL-Caltech imagery)

When Apollo 12 was orbiting the Moon, an amateur astronomer informed NASA of a transient event in Alphonsus. Dick Gordon took a look, but the crater was far to the south of his ground track and the oblique view revealed nothing unusual.

Sniffing for volcanic gases

The orbit of the Moon lies beyond Earth's magnetosphere so the Moon is usually bathed in the plasma of the solar wind. The pressure of the solar wind, however, distends the magnetosphere to form a long 'tail', through which the Moon briefly passes once per month.

To sample the solar wind, scientists provided Apollo with a sheet of exceptionally clean aluminium on a staff and the astronauts were to position it facing toward the Sun. Because this experiment was returned to Earth for analysis, the chemical composition of the solar wind could be determined more accurately than by deploying a detector on the Moon that reported its data by radio. On the other hand, the exposure time was necessarily brief, ranging from several hours to several days.

Some ALSEPs had a spectrometer to report the energy density and temporal variation of the solar wind. Some 95% of this material consists of protons and electrons, these being the products of ionising hydrogen. During the lunar night, and when the Moon was within the 'tail' of Earth's magnetosphere, the solar wind flux fell to zero.

The scientists were interested in gases that might be present at the lunar surface. An instrument called a cold-cathode ion gauge was developed to measure the pressure of such gases. At 14 orders of magnitude less than at sea level on Earth, this proved to be extremely tenuous. In effect, the exhaust from a lunar module's engine during its powered descent doubled the total amount of gas in the lunar atmosphere. The instrument was so sensitive that it could detect the presence of the astronauts walking past merely by the coolant water which sublimated from their life-support backpacks.

An instrument was deployed by Apollo 17 to identify the composition of the gas at the surface, showing it to be mostly hydrogen, helium, and neon. Interestingly, significant amounts of argon were detected at times of greater seismic activity. Since argon is the decay product of radioactive potassium it was thought that this could indicate ongoing venting by crustal faults.

On the other hand, another instrument called a suprathermal ion detector that could detect gases at the lunar surface which had been ionised by solar ultraviolet light, identified a correlation with impact rates that implied some of this gas was regolith material vaporised by meteoroids. Also, occasional trace amounts of methane, ammonia, carbon dioxide, and water hinted at the impact of small cometary fragments.

Because the escape velocity of the Moon is just 2.4km/sec, all except the heaviest of gases can readily escape and once they are ionised by solar ultraviolet they are 'picked up' by the magnetic field of the solar wind and swept away.

Dark mantles

As planning got underway for the Apollo 14 mission, it seemed that the target would be a 'dark mantle' in the eastern portion of the Sea of Serenity. This was widely believed to be a pyroclastic blanket that erupted from nearby fissures, perhaps in geologically recent times.

BELOW The original target for Apollo 14 was the 'dark mantle' in the eastern part of the Sea of Serenity. Also shown are the sites explored by Apollo 17 and Lunokhod 2. (Harland using NASA/GSFC/ Arizona State University imagery)

APOLLO 17 MISSION

Launched on 7 December 1972, it entered lunar orbit on 10 December. LM 'Challenger' landed at Taurus-Littrow on 11 December at 20.188°N, 3.774°E. Gene Cernan and Jack Schmitt performed three excursions totalling 22hr 5min, driving 35.0km, then returned to orbit on 14 December to rendezvous with Ron Evans aboard the CSM 'America', after which the LM ascent stage was crashed on a shallow trajectory 15km from the landing site on 15 December at 19.96°N, 30.50°E to provide a seismic signal. The CSM departed on 16 December and returned to Earth with 110kg of lunar material on 19 December. The S-IVB stage was crashed on 10 December on a steep trajectory at 4.3°S, 12.35°W to provide a seismic signal.

ABOVE LEFT A view of the valley nestled between massifs of the Taurus Mountains taken from the Apollo 17 lunar module shortly prior to its descent. The mothership can also be seen. The horizon is the Sea of Serenity. *(Harland using NASA imagery)*

In the event, when Apollo 13 was unable to reach the Fra Mauro Formation this target was inherited by the next mission.

When training for Apollo 15, Al Worden, who would remain in orbit around the Moon while his crewmates visited the surface, was flown over a variety of volcanic terrains on Earth in a light aircraft to assist him in his search for evidence of recent volcanism while surveying the Moon from orbit. Worden's orbits took him over a valley in the Taurus Mountains that form part of the rim of the Serenitatis basin. It was largely

LEFT A view of Apollo 17's LRV from a position on the lower slope of a massif. The boulders almost certainly rolled down from outcrops further up the mountain, beyond the reach of the astronauts, and as such were the primary sampling objective. *(Harland using NASA imagery)*

due to his reports of a dark mantling and halo craters there that led to this site being assigned to Apollo 17 as the grand finale of the program.

Gene Cernan and Jack Schmitt drove far and wide across this valley, sampling a number of objectives and collecting a record weight of samples.

The excitement came when an 'orange soil' was found incorporated into the rim of the dark-halo crater named Shorty.

To Schmitt, the only professional geologist to reach the Moon, this situation was suggestive of the kind of oxidation that occurs when gases are released through the vent of a fumarole. If that were the case, Shorty would be a volcanic crater rather than an impact crater. However, laboratory analysis proved otherwise. The orange material consisted of glass spherules issued by an ancient fire fountain.

The inferred sequence of events was that the Taurus Mountains were thrust up by the impact that made the Serenitatis basin. Later, a lava flow from the Sea of Serenity flooded the valley. Then a fire fountain laid down a blanket of pyroclastic materials. The magma that fed the fountain was rich in ilmenite, a mineral that incorporates titanium and iron oxide. If it cooled rapidly, it produced the glassy spherules

which were orange owing of the ratio of iron-to-titanium (just as the spherules at Hadley-Apennine were green because of their high magnesium content). If the droplets cooled sufficiently slowly they formed tiny crystals that were black. The area around Shorty was soon covered by a shallow flow that protected the pyroclastics until a recent impact punched through to make Shorty. Some of the excavated orange material became incorporated into the rim and the darker material was scattered to form the 'halo'.

A surprise in the Central Highlands

In compiling a geological map of a location on Earth, geologists undertake field studies of a variety of points and then generalise from this detail. But in the case of the Moon, for which in the early days only telescopic observations or images from spacecraft were available, geologists had to operate in reverse by interpreting the character of the surface in terms of the various processes that might have produced it. This was subjective and open to debate. The Apollo crews were being sent to find 'ground truth' to resolve such uncertainties.

ABOVE Jack Schmitt by the LRV on the rim of Shorty Crater. It was while inspecting a large eroded boulder on the rim that he discovered a deposit of 'orange soil'. *(NASA/Woods)*

APOLLO 16 MISSION

Launched on 16 April 1972, it entered lunar orbit on 19 April. LM 'Orion' landed in the Central Highlands on 21 April at 8.973°S, 15.498°E. John Young and Charlie Duke performed three excursions totalling 20hr 12min, driving 27.0km, then returned to orbit on 24 April to rendezvous with Ken Mattingly aboard the CSM 'Casper', after which the LM ascent stage was discarded in orbit. The CSM departed on 25 April and returned to Earth with 94kg of lunar material on 27 April. The S-IVB stage was crashed on 19 April on a steep trajectory at 1.8°N, 23.3°W to provide a seismic signal.

BELOW John Young and the Apollo 16 LRV. By this point in the program, moonwalkers were admirably equipped for field geology: observe the cuff-checklist, the shin-pocket for the hammer (absent), the chest-mounted Hasselblad and dispenser for sample bags, and the sample carrier on the side of the backpack; and tools carried on the vehicle. In the foreground is the gnomon (minus its pendulum) to document the orientation of rock samples. *(Harland using NASA imagery)*

It was widely believed that there must have been extensive volcanism in the lunar highlands. Unlike the low-viscosity magma from the mantle that rose through the crust to form the flat maria plains, the magma in the highlands would, it was argued, have formed irregular plains and lumpy hills.

Apollo 16 was dispatched to sample such a site near the large crater Descartes. As it would be another multiple-feature site, the plan was to land on a narrow plain within a cluster of hills in order to sample both types of material, at least one of which was confidently expected to be of volcanic origin.

The first thing that John Young and Charlie Duke did upon landing was to peer through their windows at the nearby rocks and, to their surprise, they saw lots of black and white ones suggestive of breccias.

By the time they had finished their surface excursions, it was apparent that the entire site

was lacking in crystalline rocks of volcanic origin.

As Ken Mattingly, in orbit, wryly remarked upon hearing this news, the scientists would have to return to their drawing boards.

With this 'ground truth', it was realised that the plain was basin ejecta which was so finely fragmented it had sloshed like a fluid and transformed the valleys into rolling plains. These rocks represent three types of breccia: regolith breccias, fragmental breccias, and impact-melt breccias – with the chemical composition of the impact-melt type resembling similar rocks from Fra Mauro, a formation which had originated in a similar manner.

The age determination of 3.76 billion years ruled out Imbrium as the source of this ejecta. It could have originated from Orientale, but this was merely an educated guess. Although most of this vast structure lies just beyond the western limb, it would have sprayed ejecta across much of the near side.

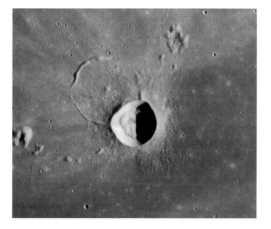

LEFT A Lunar Orbiter 4 high resolution image of the crater Lichtenberg in the Ocean of Storms which suggests that the Moon is not yet volcanically inert. *(NASA)*

Neither was there evidence to indicate a volcanic origin for the hills. By the principle of superposition, it was evident that the hills formed prior to the plain. An age of 3.92 billion years for some samples hinted that the material of the hills was ejecta from yet another basin, possibly Nectaris to the southeast, but again the association was weak.

LEFT Apollo 15 astronaut Dave Scott is bending to lift a battery powered drill during the preparation of an experiment to measure the flow of heat in and out through the lunar surface during the fortnight-long day. The rod inclined on the tool rack was used to insert a string of thermal probes into the hole. *(Harland using NASA imagery)*

141

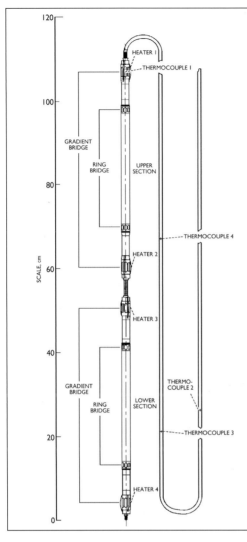

RIGHT A diagram of the string of thermal probes for the heat flow experiment from the *Apollo 15 Preliminary Science Report*. *(NASA/Marcus Langseth/Woods)*

BELOW The heat flow experiment installed by Apollo 17. *(NASA/Marcus Langseth/Woods)*

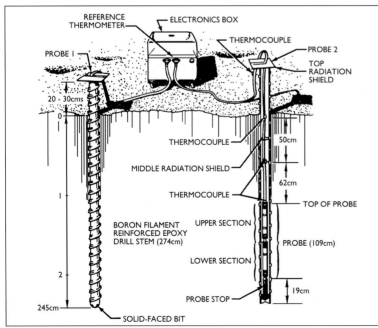

So, contrary to all expectations, Apollo didn't find any evidence of recent volcanism occurring on the Moon.

Nevertheless, the investigation continues to this day with examination of pictures taken from orbit, with encouraging results.

One good candidate is the crater Lichtenberg in the Ocean of Storms. The fact that it possesses bright 'rays' suggests the impact was 'recent', which in lunar terms means sometime within the last billion years.

The sequence of events is clear: the Ocean of Storms swamped an old crater, then Lichtenberg hit the rim of this 'ring' and sprayed a system of bright 'rays', some of which were later masked by a flow of darker lava.

The lunar heat engine

An important property of a planetary body is the rate at which it loses heat to space. Heat can come from a molten core and from long-lived radio-isotopes in the mantle. Heat loss is related to the rate of internal heat production and also to the internal temperature profile.

When Apollo 15 set down on the Sea of Rains at the foot of the Apennine Mountains, Dave Scott was to use a power tool to drive two hollow tubes into the ground to accommodate strings of thermometers.

Unfortunately this task was made arduous by a flaw in the design of the joints that linked the sections of tube. The loose material was fed up a helical exterior flute, but instead of passing over the joints it tended to clog.

Also the regolith proved to be very firm below a depth of 40cm, with the result that the sensors could be inserted only to a depth of 1.5m instead of 3m. Nevertheless, good data was received.

After the hardware had been revised, an attempt to emplace sensors in the highlands by Apollo 16 was frustrated when one of the astronauts tripped over a cable, ripping it out of the control package. Fortunately, Apollo 17 installed its sensors to the desired depth.

The airless lunar surface is a harsh thermal environment, being baked when the Sun is in the sky and frozen when it is not.

The principal investigator for this experiment

was Marcus Langseth of the Lamont-Doherty Geological Observatory at Columbia University.

The data established that in sunlight, the temperature at the top of the regolith at the mid-northern latitude visited by Apollo 15 rose to 380K. There was a net heat flow *into* the surface from solar irradiation during the day. Then at sunset the temperature fell to 100K, and during the long lunar night heat was radiated to space.

The thermal conductivity of the regolith was strongly dependent on the temperature. In fact, because the efficiency of the radiative transfer process between the fine powdery particles was proportional to the cube of the temperature, the heat flowed more readily into the ground during the day than it was lost during the night, with the result that the regolith strongly inhibited the leakage of heat from the Moon to space.

In fact, the regolith so efficiently damped the thermal variation across the lunar cycle that the temperature at a depth of 50cm was a constant 220K, and at double that depth the temperature was about 40° warmer owing to heat leaking from the interior. But this trend of the temperature increasing with depth didn't extend very deep, it was merely a shallow 'warm zone' which stored the heat that was efficiently conducting through the bedrock. An outcrop of bedrock would readily radiate its heat to space, and appear to 'glow' in a thermal image.

Scott also used a power tool to drive a multi-segmented core tube to a depth of 2.4m, a little short of the desired depth. But retrieving the tube required the combined toil of the two men. And owing to an incorrectly assembled tool, they had difficulty separating the tube into its segments for stowage. Obtaining this sample consumed much more time than had been allocated but the results were worth it.

When the core was examined afterwards it proved to comprise at least 42 distinct layers of material. The uppermost 45cm was heavily intermixed by the meteoritic 'gardening' process, but below this it was unsorted and variations in the chemical composition told a story.

Dating established that the top of the unsorted section was 400 million years old. This meant that activity on the plain during this period of time was confined to the uppermost 50cm of the regolith. On Earth during this interval, continents formed, then split apart, drifted thousands of kilometres, and collided to produce mountain ranges. In general therefore, the lunar surface is truly timeless.

Soviet rovers

As Scott and Irwin drove around Hadley-Apennine using their LRV, theirs was not the only vehicle on the Moon.

In November 1970 the Soviets had landed Luna 17 near the Bay of Rainbows, on the

LEFT A depiction of Lunokhod 1 ready to drive off its Luna 17 landing stage onto the lunar surface. *(Academy of Sciences of the USSR)*

northern shore of the Sea of Rains. After ramps were unfolded, an eight-wheeled vehicle drove down to the surface.

This 'Lunokhod' had cameras like those used by Luna 9 for panoramic views and a pair of TV cameras on the front to provide stereoscopic vision for its operators on Earth.

It immediately raised a flap to expose a solar panel to charge its battery, and then closed up for its first lunar night.

After sunrise it started driving south, pausing occasionally to measure the bearing strength of the surface with a penetrometer and to study the chemistry of the regolith, finding it to be basaltic. It parked for its second night about 1.4km away from the lander, returned by a roundabout route the following day, then set off north, never to return. When it expired in October 1971 it had travelled a total of 10.5km, provided 200 panoramas, and analysed the regolith at 25 sites.

A few weeks after Apollo 17 wrapped up the first phase of America's exploration of the Moon, Luna 21 landed inside the crater Le Monnier on the eastern shoreline of the Sea of Serenity and deployed a second Lunokhod, which set off in a southerly direction. Upon encountering a canyon that was 300m wide it drove several kilometres along the rim, dodging boulders. Unfortunately, it suffered thermal problems and failed after just three months having covered a total distance of 42km.

Although the Soviet capability to robotically return samples of lunar material to Earth and to drive across its surface established that work could be undertaken without sending humans, the Apollo crews carried out genuine geological field trips and returned with much larger hauls of samples that had been carefully selected *in situ*.

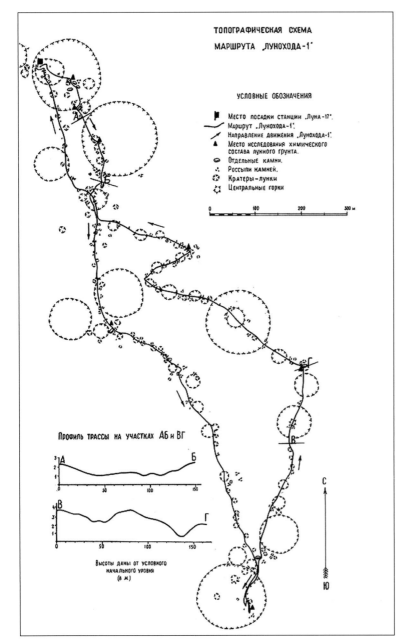

LUNOKHOD MISSIONS

Luna 17
Launched on 10 November 1970, it entered lunar orbit on 15 November and landed in a northern bay of the Sea of Rains on 17 November at 38.28°N, 35°W. It deployed the Lunokhod 1 rover which, when contact was lost in September 1971, had driven a total distance of 10.5km.

Luna 21
Launched on 8 January 1973, it entered lunar orbit on 12 January and landed in the partially flooded crater Le Monnier on the eastern shore of the Sea of Serenity on 15 January at 25.85°N, 30.45°E. It deployed the Lunokhod 2 rover which, when contact was lost in May 1973, had driven a total distance of 42km.

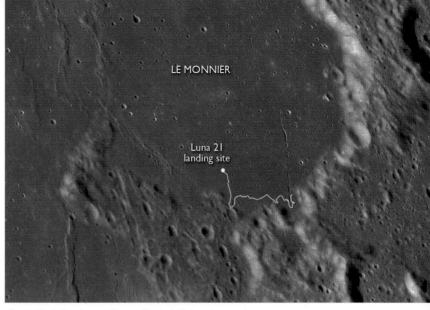

LE MONNIER

Luna 21
landing site

The origin of the Moon

Impressed by the rings of Saturn, the French mathematician Pierre-Simon Laplace proposed in 1796 that the solar system was created during the gravitational collapse of an enormous cloud of gas which was in a state of rotation. This analysis was prompted by what he considered to be several remarkable facts (as far as was known at that time):

(1) All the planets and moons travel around the Sun in the same direction.
(2) All the planets rotate on their axes in that direction.
(3) Apart from minor departures, the planets all travel around the Sun in the same plane.

Laplace reasoned that as the cloud of gas was shrinking, the conservation of angular momentum would have made its rate of rotation increase. Furthermore, he argued, it would have repeatedly shed material in order to relieve itself of 'excess' angular momentum, thereby making a series of concentric rings in a single plane. After the central mass had become the Sun, each ring would have condensed to create a planet in a near-circular orbit at that particular distance from the Sun.

In likewise manner, the process of planetary formation would have shed local rings that became satellites; in Earth's case, the Moon.

Although this *nebular hypothesis* was widely accepted at the time, further mathematical study showed it wouldn't work as Laplace had imagined.

George Darwin posited in 1878 that Earth and the Moon were formed together. In this scenario, the initial rapidly rotating body of hot fluid became an ellipsoid that spun about its minor axis in an unstable equilibrium with two forces acting on it; namely its own natural period of vibration and the gravitational tides raised by the Sun. Once these forces had achieved a resonance, the shape would become ever more like a dumbbell until the narrow 'neck' broke, whereupon, Darwin argued, the larger portion became Earth and the smaller portion the Moon.

This *fission hypothesis* was also ultimately discarded due to mathematical deficiencies, not the least of which was that a spinning dumbbell of fluid would tend to divide into two more or less similar masses.

As the Space Age began, Harold Urey and Gerard Kuiper argued about the origin of the Moon. The main point of dispute was whether the interior was 'pristine' material which was representative of the solar nebula and therefore incapable of volcanic activity or had

HERE MAN COMPLETED HIS FIRST
EXPLORATIONS OF THE MOON
DECEMBER 1972, A.D.
MAY THE SPIRIT OF PEACE IN WHICH WE CAME
BE REFLECTED IN THE LIVES OF ALL MANKIND

EUGENE A. CERNAN
ASTRONAUT

RONALD E. EVANS
ASTRONAUT

HARRISON H. SCHMITT
ASTRONAUT

RICHARD NIXON
PRESIDENT, UNITED STATES OF AMERICA

ABOVE The inferred traverse route of Lunokhod 2 in the breached crater Le Monnier. *(Harland using data by the Academy of Sciences of the USSR and imagery from NASA/GSFC/Arizona State University)*

LEFT The design of the plaque on the descent stage of the Apollo 17 lunar module. *(NASA)*

APOLLO MISSIONS' SURFACE STATISTICS

	Distance (km)	Experiments (kg)	EVA (hr)	Samples (kg)
Apollo 11	0.25	102	2:24	21.55
Apollo 12	2.0	166	7:29	34.35
Apollo 14	3.3	209	9:23	42.28
Apollo 15	27.9	550	18:33	77.31
Apollo 16	27.0	563	20:12	95.71
Apollo 17	35.0	514	22:05	110.52
TOTAL	95.45	2104	80:06	381.72

undergone the thermal differentiation required to facilitate volcanism.

Urey argued that the 'cold Moon' had formed independently elsewhere in the solar system and later been captured by Earth.

Kuiper reasoned the Moon formed in the same region of the nebula as Earth, with the two becoming gravitationally bound early on.

To general consternation, the Apollo samples ruled out *all* of these theories!

It wasn't until 1984 that a consensus was arrived at in which the Moon was formed by the accretion of debris left over from the collision of a Mars-sized body with the proto-Earth very soon after they had coalesced from the solar nebula. This was wryly nicknamed the *Big Splat hypothesis*.

Ever since the Moon coalesced in orbit around Earth, it has been receding due to tidal effects.

It was evident to the Greek explorer Pytheas circa 300 BC that the tides were correlated with the Moon but the actual relationship remained a mystery until Laplace explained it using Newton's gravitation.

In fact, Laplace found that although the main

influence was the Moon, there was a secondary effect by the Sun.

The gravitational pull of the Moon is fractionally stronger on the side of Earth that faces the Moon than it is on the opposite side. This is due to the gradual weakening of a body's gravity with distance. The net effect is to impart a slight stretching force across the planet. As a result, the Moon raises a bulge of just over half a metre in the open ocean on either side of Earth. The Sun has a similar effect, but its greater distance reduces its significance to about a quarter that of the Moon.

Because Earth is rotating on its axis much more rapidly than the Moon travels around its orbit and the Moon migrates eastward relative to the stars, Earth must rotate a little more than 360° in order to catch up with the line to the Moon. High tides therefore occur at intervals related to the time between lunar transits of the local meridian.

When the Sun, Moon and Earth are aligned, the overall tidal effect is reinforced to produce an especially high tide named a *spring tide*. This only occurs when the Moon is at its 'new' or 'full' phase. When the Sun and Moon are at 90° with respect to Earth, the solar tidal bulges partially cancel the Moon's effect, creating a *neap tide*.

A tide is of little significance in the open ocean, the ebb and flow is essentially a coastal effect. To a first approximation the bulges remain aligned near the lunar meridian and the shore encounters them on a twice-daily basis. In fact, the high tide that occurs beneath the Moon is slightly larger than that in the opposite hemisphere.

In addition to raising bulges in the ocean, the Moon's gravitational attraction causes Earth's crust to flex upward by almost 25cm (although we don't notice this in everyday life). Unlike the bulges of water, which remain close to the lunar meridian, the crustal bulges are slow to rise and slow to relax.

The fact that these bulges are always 'ahead' of the Moon imparts a 'lever arm' which accelerates it in its orbit. This is distinct from the radial acceleration towards Earth's centre that, as Newton realised, causes the Moon to orbit Earth instead of flying off into space.

In the same way that the bulge on the rapidly rotating Earth accelerates the Moon in its orbit, the Moon slows Earth's rate of rotation. Over time, the Moon is accelerating and hence receding (lengthening the month) and the axial rotation of the planet is slowing down (lengthening the day by about 1sec per 100,000 years).

A planar array of reflectors was placed on the Moon by Apollo 11 and two later missions, and also carried on the two Lunokhod vehicles.

These high-precision corner-cube reflectors were able to redirect a laser beam from Earth back to its source. Although divergence of the beam in transit to the Moon meant it illuminated a broad footprint that diluted the energy impinging on the instrument and the return beam was further diluted, it was possible to detect the signal. The transit time of each pulse gave an instantaneous measurement of the line-of-sight distance between the telescope which fired the laser and the reflector on the Moon to an unprecedented accuracy.

Prior to the installation of these reflectors,

ABOVE The laser reflector emplaced on the Moon by Apollo 11. *(NASA)*

the distance to the Moon was measurable only to an accuracy of 100m. In the 1970s the reflectors made it possible to measure the distance to a given point to within about 20cm and today's technology can achieve an accuracy of better than 1cm.

Long-term monitoring has established that the transfer of angular momentum from Earth to the Moon is causing the Moon to recede at an average rate of 3.8cm per annum.

This implies that when the Earth–Moon system formed some 4.5 billion years ago, the planet was rotating much more rapidly and its satellite was much closer in. In fact, over time the orbital angular momentum of the Moon has come to exceed the rotational momentum of its primary.

Just as the Moon flexes Earth's crust, Earth's much stronger gravity raises a bulge in the Moon. With the passage of time, the 'drag' of this bulge has obliged the axial rotation of the Moon to synchronise with the period of its orbit and thereby maintain a single hemisphere facing Earth. As the Moon continues to recede, the tidal forces will further slow its spin to maintain this synchronicity.

As George Darwin wrote in his 1898 book *The Tides and Kindred Phenomena in the Solar System*, the transfer of angular momentum will continue until the axial rotation of Earth is synchronised with the orbital period of the Moon, at which time the two bodies will maintain the same hemispheres facing each other. He calculated the critical value as 47 days (in current terms). However, this will not occur for another 50 billion years. To put this

into context, the universe itself originated in the Big Bang only 14.7 billion years ago and the Sun formed about 5.5 billion years ago.

Scientists had inferred from the seismometers installed by the Apollo crews that the Moon had a crust, a mantle, and a core. The measurements of how the Moon is retreating from Earth gave information about the distribution of internal mass.

The Moon is now believed to possess a solid iron-rich inner core and a fluid outer core primarily of liquid iron. Surrounding this is, in turn, a partially molten boundary layer, the mantle, and the crust.

The core is roughly 20% of the mean radius of 1,740km whereas the terrestrial core occupies 50% of Earth's radius of 6,370km. The mantle consists mainly of the relatively dense minerals olivine and pyroxene. It formed early in the Moon's history, when the surface was a deep magma ocean. As with Earth, the mantle and crust are separated by a well-defined boundary. The crust is rich in the lightweight mineral plagioclase, and its average thickness of 70km is thrice that of the terrestrial continental crust.

Once the motions of the Moon were understood, it was possible to study the terrestrial contributions in the laser data in order to study variations in Earth's rotation rate and precession of its axis. Indeed, it became possible to directly measure the rates at which the process of plate tectonics is driving continents around the globe.

Eclipses

The first complete astronomical record of a solar eclipse was made by the Babylonians, who noted its start and end times on 19 March in 721 BC. They were meticulous observers of celestial events, and interpreted them in terms of an elaborate astrology. Their observations led to the discovery of the Saros cycle, in which lunar eclipses obey a cycle with a period of about 223 synodic months, or approximately 18 years.

The ability to predict when a lunar eclipse would occur would have been regarded by the uninitiated as a powerful display of astrological prowess.

However, whereas a lunar eclipse is visible from anywhere on the night-time hemisphere of

Earth, any given solar eclipse can be seen only from a certain part of the globe, therefore the Babylonians would have had little success in predicting solar eclipses.

That a solar eclipse happens when the Moon passes in front of the Sun was realised by Anaxagoras of Athens in the 5th century BC. The Greek statesman Pericles, who was familiar with this, explained it to the captain of a ship who had refused to set sail in the aftermath of an eclipse on 3 August 431 BC, lest it be a bad omen. Satisfied that the phenomenon was natural, the captain sailed and the journey was uneventful.

Aristarchus, who was born on the island of Samos in 310 BC, inferred from the fact that the Moon could completely mask the Sun during a solar eclipse that their angular diameters were comparable and estimated the size at half of a degree (unfortunately, some of his successors incorrectly wrote that his estimate was all of 2°).

From the circular profile of the shadow that Earth casts on the Moon during a lunar eclipse, Aristarchus estimated the shadow to possess a diameter 3.8 times the size of the Moon's disc.

Next, he conceived of a way to measure the relative distances of the Sun and Moon. When the phase of the Moon is either 'first quarter' or 'last quarter', showing half of its face, the angle between an observer, the Moon, and the Sun will be 90°. Aristarchus reasoned that

ARISTARCHUS'S LUNAR DISTANCE

By elementary trigonometry the diameter of the Moon is the product of its distance and sin(θ), where θ is its angular diameter. The distance is therefore the diameter divided by the angle. For Aristarchus's estimate of the apparent diameter as 0.5° and a ratio of 3.8 for the diameters of Earth and the Moon, he was able to calculate as follows: Sin(0.5°) = 0.008, the diameter of the Moon is 0.263 (this being the inverse of 3.8) and therefore 0.263/0.008 is approximately 30 for the distance to the Moon in terms of Earth's diameter.

Crater · Mantle · Crust · Partial melt · 480km · Outer core (liquid) · 330km · Inner core (solid) · 240km · 1,737km · 1,685km · Lunar terra · Lunar maria

ABOVE The internal structure of the Moon inferred from modern data. *(Woods)*

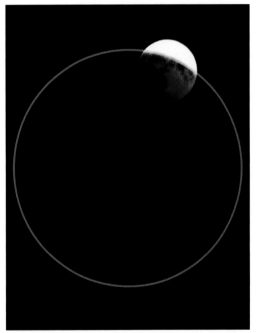

LEFT A recreation of the Moon entering the Earth's shadow during a lunar eclipse, with Aristarchus's estimate of the shadow being 3.8 times the diameter of the lunar disc. *(Harland/Woods)*

BELOW Aristarchus's method for calculating the relative distances of the Sun and Moon. *(Woods)*

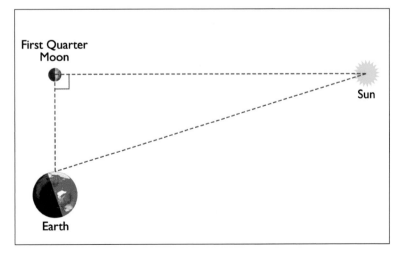

First Quarter Moon · Sun · Earth

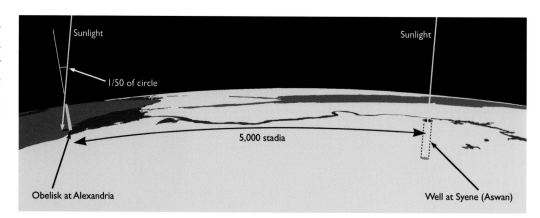

RIGHT How Eratosthenes was able to use a solar eclipse to measure the diameter of Earth. *(Woods)*

Sunlight

1/50 of circle

Sunlight

5,000 stadia

Obelisk at Alexandria

Well at Syene (Aswan)

measuring the broad angle to the Sun relative to the centre of the lunar disc would allow him to calculate the narrow angle between an observer, the Sun, and the Moon. His estimate of 3° implied the Sun was 20 times more distant than the Moon.

Although his technique was valid in principle, it was difficult for a naked-eye observer and Aristarchus's value was far short of the true ratio of 400. Nevertheless, this was the first indication of the relative scale of objects in the celestial realm.

Reasoning that because the Sun was so far away, the diameter of Earth's shadow at lunar distance would be only slightly diminished, and knowing the size of the Moon's disc in the sky and its size relative to Earth's diameter, Aristarchus calculated the distance to the Moon to be 30 times the diameter of Earth.

This result would remain a relative measure until someone was able to measure the size of Earth.

In the 3rd century BC Eratosthenes, the chief librarian in Alexandria, realised that he had the means to measure the size of Earth.

At summer solstice, the position of the Sun reached its highest possible elevation in the sky. At noon it was directly overhead at Syene (now Aswan) and it shone directly on the water at the bottom of a deep well. At the same time in Alexandria, some distance to the north along the same meridian, the Sun was 7.2° from the zenith.

Knowing the distance between the sites, Eratosthenes used this angle to calculate the circumference of Earth and hence its diameter.

The unit of distance in those days was the stadion. We aren't certain how the distance of 5,000 stadia cited by Eratosthenes converts

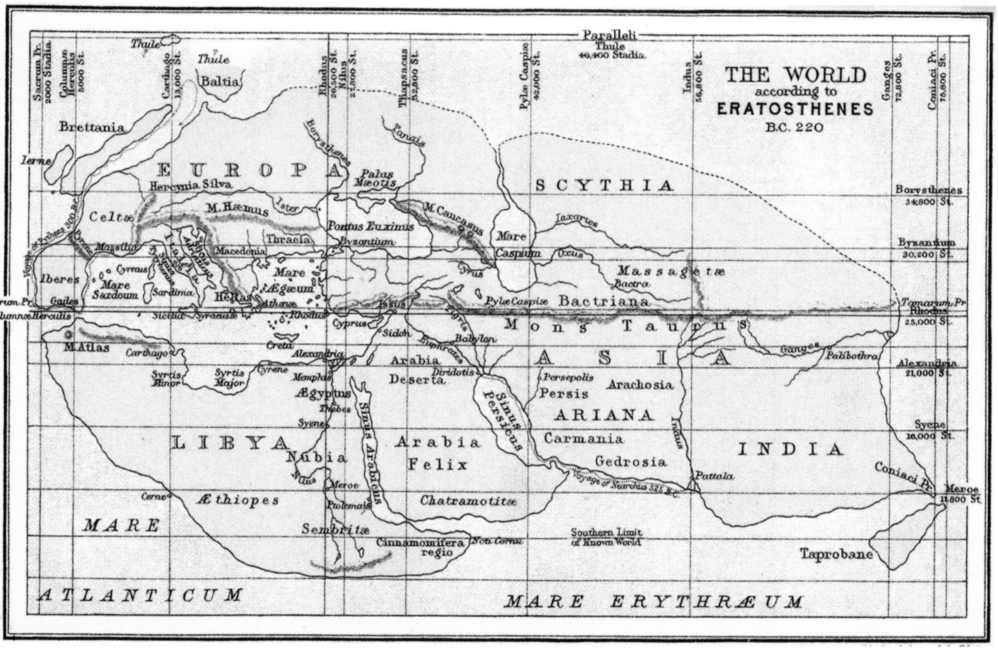

RIGHT Eratosthenes's announcement that Earth was an enormous sphere wasn't well received because it would imply that the 'known world', shown here in this 19th-century reconstruction, was just a tiny portion of its surface. *(Heritage History)*

THE WORLD according to ERATOSTHENES B.C. 220

to modern units (the actual distance is 787km)
but he achieved a realistic sense of the size
of Earth. In fact, his value was received with
considerable scepticism at the time since it
implied the 'known world' to be merely a tiny
fraction of a vast globe.

Nevertheless, by estimating Earth's size
Eratosthenes provided a scale for Aristarchus's
sense of the true distances to the Moon and to
the Sun.

In the 2nd century BC, Hipparchus realised
the locations of two observers provided the
basis for a parallax measurement that could be
used in trigonometry. He examined records of
solar eclipses seen from Alexandria and Nicaea,
situated on the same meridian and separated
by a known distance, then used the extents to
which the Moon had masked the solar disc to
calculate the distance to the Moon as being 59
to 67 times the radius of Earth. This confirmed
Aristarchus's estimate.

The apparent diameter of the Moon varies
inversely with its distance from Earth, being
largest when at perigee and smallest when at
apogee. The maximum and minimum diameters
also vary as the perigee and apogee distances
vary. In fact, the disc varies between 29.3 to
34.1arc-min. Its mean apparent diameter is
currently 32arc-min. In addition, the diameter
of the solar disc varies slightly as a result of the
eccentricity of Earth's orbit.

If the Moon crosses the ecliptic precisely
on the line that joins Earth to the Sun then in
certain situations the conical shadow of the
Moon is able to cast a small spot of darkness
onto the surface of Earth to create a total
eclipse.

If an eclipse occurs when Earth is at aphelion
with the apparent diameter of the Sun at its
minimum, and the Moon is at perigee with its
apparent diameter at its maximum, then the
size of the shadow spot will be at its largest,
spanning 269km. However, if the eclipse occurs
when Earth is at perihelion and the Moon is at
apogee, then the shadow cone won't reach
Earth and the result is an annular eclipse in

RIGHT A map showing the track of the solar eclipse of 11 August 1999. *(Woods/Map from National Geographic's MapMaker Interactive)*

BELOW The red hue of the Moon passing through the Earth's shadow as it rises during a lunar eclipse. *(Woods)*

which a ring of the solar disc remains visible around the lunar disc.

The motions of the Sun and the Moon in the sky and the rotation of Earth on its axis combine to make the shadow spot trace out the *track of totality*. In addition to these factors, the duration of totality depends on the observer's latitude. Taking into account all pertinent factors the longest possible duration of totality is 7.5min.

The fact that Earth is much larger than the Moon means its shadow cone is broader and longer. Indeed, as Aristarchus surmised, it extends far beyond the orbit of the Moon. If the Moon grazes the shadow cone, the lunar disc becomes slightly dimmed across one limb. If the Moon passes directly through the shadow, the eclipse might last 100min.

Even when the Moon passes through the centre of Earth's shadow, it remains faintly visible. Kepler was the first to explain this illumination as sunlight passing through Earth's atmosphere. Isaac Newton explained its reddish hue in terms of refraction.

The sky appears blue in daylight because the blue portion of the spectrum of light is preferentially scattered by the molecules which comprise air. At the time of a lunar eclipse, the Sun is low on the horizon on Earth's circular limb and some of that light passes straight through and is refracted into the shadow cone. But having passed through a large amount of Earth's atmosphere, the light that is refracted into the cone is preferentially the red portion of the spectrum.

This refraction is more pronounced in some cases than in others because the amount of light that is fed into the shadow cone depends on the cloudiness of the atmosphere in the limb regions and the amount of atmospheric dust.

Because the mean radius of the Moon's orbit is progressively increasing, in the past the lunar disc would have been larger than the solar disc, and in the future it will be smaller, at which time total solar eclipses will cease.

RIGHT When the Apollo 12 astronauts were returning to Earth their spacecraft passed through the Earth's shadow. *(NASA)*

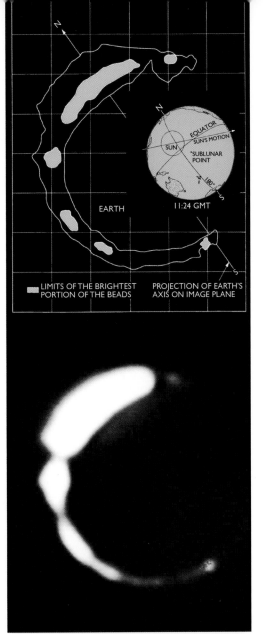

LEFT The Surveyor 3 lander observed the Earth eclipsing the Sun on 24 April 1967. The graphic explains the geometry and how the atmosphere refracted sunlight. *(NASA)*

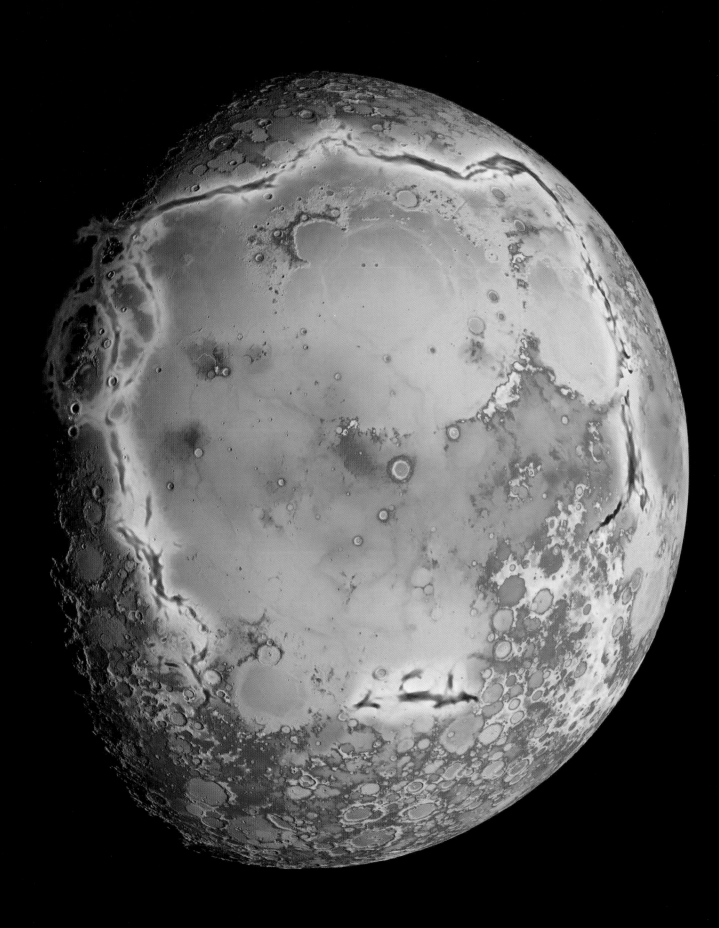

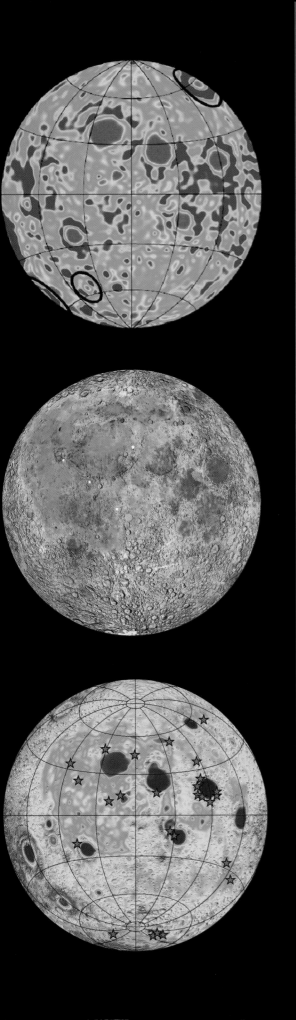

Chapter Ten

The modern era

This final chapter explains
the discoveries made by the
resurgence of interest in the Moon
since the 1990s, with many nations
participating both singly and jointly
in a new 'golden age'.

OPPOSITE Modern instruments reveal the Moon in
unprecedented detail. Left: the gravity gradient data from
the GRAIL spacecraft displayed on a topographic map from
the laser altimeter on Lunar Reconnaissance Orbiter. *(NASA/
Colorado School of Mines/MIT/GSFC/Scientific Visualization
Studio)*. **Top: the gravity anomalies mapped by Lunar
Prospector.** *(NASA)* **Centre: thorium abundance mapped by
Lunar Prospector.** *(NASA/JPL-Caltech/Jeff Gillis with thanks to
Paul Spudis of the Lunar and Planetary Institute)*. **Bottom: crustal
thickness inferred from the GRAIL data and topography
mapped by LRO. The star symbols represent the olivine-rich
materials mapped by the Japanese Kaguya mission.** *(NASA/
JPL-Caltech/IPGP)*

After the final Apollo lunar landing, NASA lost interest in the Moon and the Soviet Union soon followed suit. It wasn't until 1994 that the United States sent another vehicle to the Moon and this time it was a technology demonstrator funded by the Department of Defense.

For the last three Apollo missions a scientific instrument module had been carried in a vacant bay of the spacecraft that stayed in orbit while the lunar module conducted the surface expedition. In addition to high resolution mapping, these payloads studied the composition of the surface. The results were intriguing but the remote sensing technology was primitive.

Clementine

The Clementine spacecraft was launched to test the latest lightweight, miniaturised sensors for the Strategic Defense Initiative. The military had joined with NASA to map the Moon prior to departing in order to intercept an asteroid scheduled to make a close pass by Earth.

On 19 February 1994 the spacecraft entered into an orbit that was perpendicular to the lunar equator and varied in altitude between 430 and 2,950km, with the low point in the mid-southern hemisphere.

While near perilune, Clementine undertook laser altimetry and obtained imagery spanning the spectrum from the near-infrared through to

ABOVE The scientific instrument bay of the Apollo 15 mothership. *(NASA)*

RIGHT Detail of the scientific instrument bay. *(NASA/Woods)*

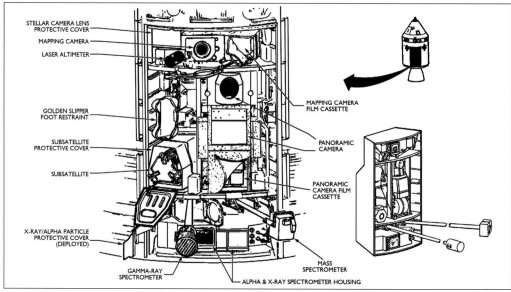

STELLAR CAMERA LENS PROTECTIVE COVER
MAPPING CAMERA
LASER ALTIMETER
MAPPING CAMERA FILM CASSETTE
GOLDEN SLIPPER FOOT RESTRAINT
PANORAMIC CAMERA
SUBSATELLITE PROTECTIVE COVER
SUBSATELLITE
PANORAMIC CAMERA FILM CASSETTE
X-RAY/ALPHA PARTICLE PROTECTIVE COVER (DEPLOYED)
MASS SPECTROMETER
GAMMA-RAY SPECTROMETER
ALPHA & X-RAY SPECTROMETER HOUSING

ultraviolet for a multispectral investigation of the composition of the surface in much more detail than was possible with Apollo.

At the apolune of the 5hr orbit, all this data was downlinked to Earth. Over the period of a month, as the Moon turned on its axis beneath the plane of the spacecraft's orbit, Clementine was able to inspect the entire range of southern longitudes. Then the point of perilune was moved to the northern hemisphere to complete the coverage.

Early evidence for water ice

For many years the lunar polar regions had fascinated scientists. The last two Lunar Orbiter missions had observed them, but only in visible light. They were far beyond the reach of the Apollo missions. Clementine provided a vertical perspective of each polar region across the full range of illumination of a lunation cycle.

Whilst Earth's rotational axis is offset 23.5° to the pole of the ecliptic, which is the plane in which Earth orbits the Sun, the axis of the Moon is inclined by a mere 1.54°. This raised the possibility of there being polar craters whose floors are never illuminated.

In 1961 Kenneth Watson, Bruce Murray and Harrison Brown, all at the California Institute of Technology in Pasadena, speculated that the floors of permanently shadowed craters might form 'cold traps' in which the temperature would remain at about 100K. In that case, water ice delivered over the aeons by comets would tend to accumulate.

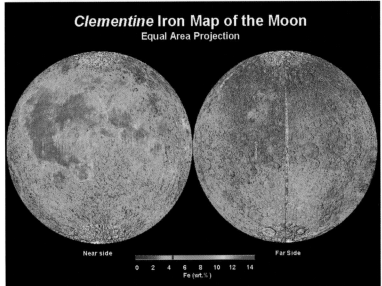

CENTRE This map of iron at the lunar surface by Clementine marks out the mare plains. (Clementine/Paul Spudis of the Lunar and Planetary Institute)

RIGHT This map of titanium at the lunar surface by Clementine marks only some of the mare plains. (Clementine/Paul Spudis of the Lunar and Planetary Institute)

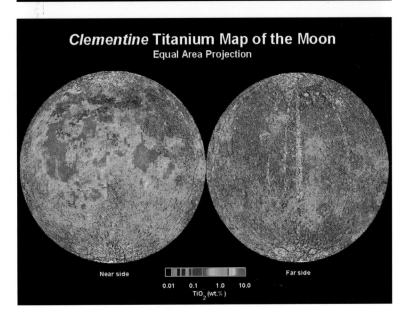

Clementine's orbit gave an opportunity for a team led by Gene Shoemaker to test this theory. The bistatic radar experiment relied on the fact that water ice has a unique 'signature' when it is probed by radio energy. It reflects most efficiently when the angle of incidence of the illumination is near-zero (coherent backscatter) and, unlike rock, ice preserves the polarisation of the wave.

The geometry for such a test was suitable in March and April 1994. Clementine 'beamed' its radio signal into the shadowed craters at each pole and NASA's antennas monitored the reflection. The March test was inconclusive. The results in April were intriguing: two passes over the north pole were unremarkable, as was one near the south pole, but as the 'beam' tracked across the shadowed areas right at the pole on orbit 234 the terrestrial antennas noted both an increase in the strength of the reflection and a greater percentage of the signal preserving the polarisation.

This data was reported as possible evidence that water ice comprised a portion of the surface layer in a permanently shadowed crater.

Clementine set off for asteroid Geographos on 3 May. Unfortunately, a few days later, while rehearsing manoeuvres which it was to perform during the 100km flyby, a computer malfunction sent the craft tumbling out of control.

Lunar Prospector

In the 1990s NASA began to send small low-cost missions into deep space, one of which was called Lunar Prospector.

This spacecraft initiated its science activities on 16 January 1998 in circular polar orbit at an altitude of 100km. In addition to investigating the localised magnetic fields, it was to characterise the mineralogy of the surface on a global basis, including seeking evidence of water ice in polar craters. The first rumour that ice had been found was leaked a month later, and on 5 March it was confirmed that the data did indeed imply the presence of a significant amount of water ice at both poles.

The neutron spectrometer instrument didn't detect water ice directly; this was inferred from the detection of hydrogen by remotely sensing neutrons. The reasoning was that when a high-energy cosmic-ray particle strikes the nucleus of an atom in the regolith, the interaction releases neutrons. Most of these neutrons are 'hot', meaning that they fly off at high speed. If such a neutron was later to encounter a lightweight nucleus it would yield some of its energy and slow down. A molecule of water contains two hydrogen atoms, so, being the lightest of atoms these would be most affected.

Scanning the surface to measure the energy spectrum of the neutrons, the instrument could identify the *epithermal* neutrons which implied the presence of hydrogen. The instrument saw significant indications at both poles.

By the end of the mission in 1999, the data was consistent with an enrichment of hydrogen caused by a layer of water ice beneath perhaps 40cm of desiccated regolith. Because the spatial resolution of the scan was only 100km, it wasn't possible to specifically identify the locations, but there was a correlation with the suspected cold traps.

However, because ice was inferred from

BELOW An artist's depiction of the Lunar Prospector spacecraft. *(NASA/Ames)*

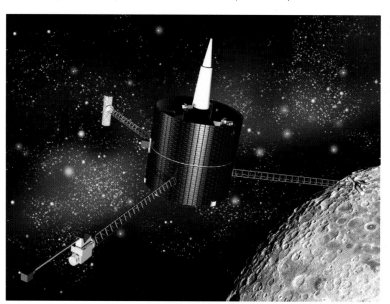

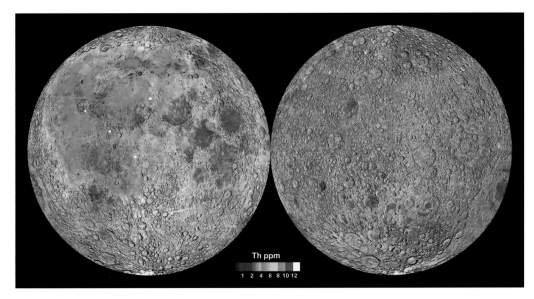

LEFT A map of thorium at the lunar surface from gamma-ray spectrometry by Lunar Prospector. Its concentration evidently correlates with the KREEPy rocks that are rich in potassium, 'rare earth' elements and phosphorus. A heterogeneous distribution at the surface implies fundamentally different geological histories of the individual lunar regions. (NASA/JPL-Caltech/Jeff Gillis with thanks to Paul Spudis of the Lunar and Planetary Institute)

the presence of hydrogen in the regolith, and the lunar surface is irradiated by the solar wind (essentially hydrogen ions), other scientists, most notably Jack Schmitt who walked on the Moon in 1972, argued that the inferred hydrogen was more likely to be present in this form.

Another of Lunar Prospector's achievements was to map the distribution of radioactive isotopes, and in particular thorium, to provide further insight into the process that made the KREEPy additive in the western maria.

As a finale, in July 1999 the spacecraft was steered to crash into a permanently shadowed crater. It was calculated that there was a 10% chance of the impact vaporising sufficient water to produce a plume capable of being detected by terrestrial observers but nothing was seen. Perhaps the vehicle struck at such a shallow angle that it was smashed to pieces without excavating to a sufficient depth.

Gene Shoemaker had lost his life in a traffic accident in 1997, so in a poignant tribute Lunar Prospector was carrying a vial of his cremated remains. To date, he is the only person to have had his ashes scattered on the Moon.

LUNAR PROSPECTOR MISSION

Launched on 7 January 1998, it entered a polar lunar orbit on 11 January to conduct a global survey with a variety of instruments.

Terrestrial radars seek water ice

When conditions are suitable, the beam from a terrestrial radio telescope operating as a radar can provide shallow-angle illumination of some of the permanently shadowed craters at the lunar poles. Radio is able to penetrate the regolith to a depth that is several times its own wavelength, so in principle a radar can 'probe' for the presence of ice.

The 305m dish of the Arecibo Observatory on the island of Puerto Rico is operated by Cornell University for the US National Astronomy and Ionosphere Center. After the results of the Clementine bistatic radar test were

BELOW The 305m dish of the Arecibo radio telescope on the island of Puerto Rico is the world's most powerful 'planetary radar'. (Arecibo Observatory)

announced, Arecibo studied the polar regions at a wavelength of 12.6cm. The results, reported in 1997, were ambiguous because whereas the strongest reflections were from sites at the south pole that could well be permanently shadowed, strong reflections were also received from areas nearby that weren't. This led the team to explain the reflections they received in terms of the roughness of the terrain, because a slope that is perpendicular to the line of sight will bounce a radio wave back towards its source.

After Lunar Prospector charted hydrogen in the polar regions, it was decided to use the Goldstone Solar System Radar in California to locate the cold traps.

This 70m dish illuminated the target zones at a wavelength of 3.5cm and the echoes were received by a pair of 34m antennas that were 20km apart. When the signals were later processed using an interferometry technique it was possible to compile a three-dimensional digital elevation model of these regions by measuring surface elevations at a network of points that were spaced out horizontally at intervals of 150m, with each vertical measurement accurate to 50m. After a computer inferred how this landscape would be illuminated by the Sun, it was announced in 1999 that parts of the floors of five large craters located in the south polar region were indeed in eternal shadow.

Although a wavelength of 3.5cm provided a high resolution for mapping purposes, it couldn't penetrate very far into the regolith in search of ice.

So the Arecibo telescope made a follow-up study that probed these interesting areas at a wavelength of 70cm to reach the layer that was envisaged by the Lunar Prospector team. When the results were published in 2003 they weren't encouraging. It wasn't possible to say that there was no ice, but the data imposed strict constraints.

For Lunar Prospector's reports of hydrogen enrichment in the upper half-metre of regolith to be consistent with the lack of radar backscatter from a depth of several metres, it was necessary to postulate there couldn't be a slab of solid ice at or near the surface – such ice as might be present at shallow depth must either be in layers that were thinner than the radar wavelength and were interbedded with desiccated regolith or be in the form of individual grains that were intermixed with the fragments of rock.

Observations were then conducted at 12.6cm at the high spatial resolution of 20m in an effort to localise specific samples. The results reported in 2006 made thin layers at shallow depth unlikely. This left individual grains of ice mixed in with the fragmental material as the only possibility. However, it would be impossible to detect those by radar.

In the meantime, D. H. Crider of the Catholic University of America in North Carolina and R. R. Vondrak of NASA's Goddard Space Flight Center in Maryland made a mathematical study of how a cold trap would evolve in response to the *space weathering* that 'gardened' the plasma delivered by the solar wind into the regolith. The results, published between 2001 and 2003, concluded that the regolith would attain a steady state of around 4,000 parts per million of water. Of course, the water hadn't been present in the plasma; it was created by hydrogen in the solar wind combining with oxygen atoms drawn from oxides in the regolith.

Importantly, the results of this analysis were consistent with the radar data.

The model suggested that the layer in which the ice grains were present increased with time and could attain a thickness of 1.6m after 1 billion years.

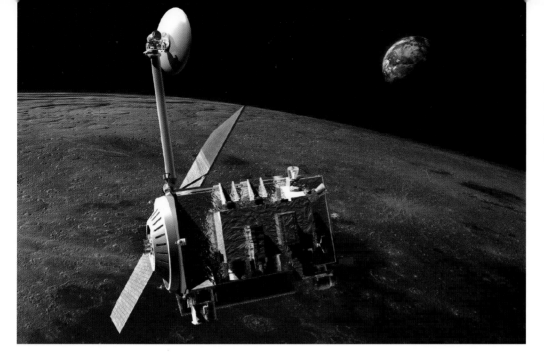

Direct evidence of water

In 2009 the NASA Lunar Reconnaissance Orbiter (LRO) began its operations in a circular polar orbit at an altitude of 50km. In addition to making a global geodetic survey and mapping mineralogical resources, it was to use several new instruments to investigate the permanently shadowed polar areas in an effort to confirm the presence of water ice. Another key task was to identify elevated areas near the poles which are always illuminated and to map such topography in order to assist in the selection of landing sites for future landers.

It soon found evidence of the presence of water. Its neutron detector could measure the distribution of hydrogen within the uppermost metre of regolith with a spatial resolution of 10km; much better than Lunar Prospector. A surprise was that it found hydrogen enrichment in many areas, not only in the permanently shadowed craters.

Another instrument could use starlight and ultraviolet sky-glow as a source of illumination to undertake optical imaging of areas that were not illuminated by sunlight, enabling it to map areas in shadow. This detected absorption at 1,600 angstroms (160nm) in

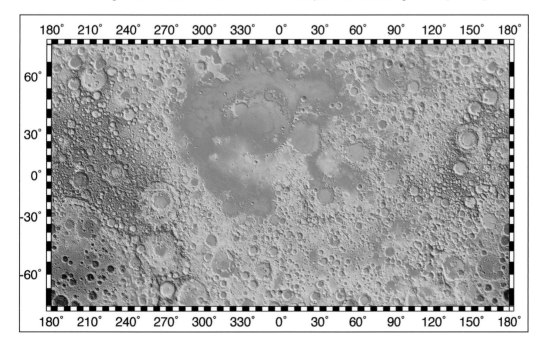

LEFT A topographic map of the Moon (excluding the polar regions) produced by the laser altimeter on LRO. *(NASA)*

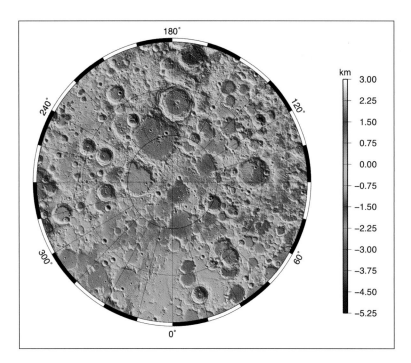

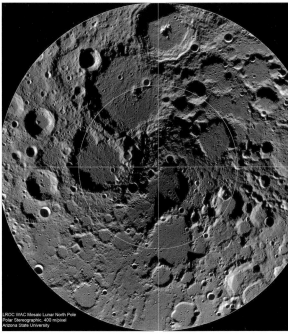

LROC WAC Mosaic Lunar North Pole
Polar Stereographic, 400 m/pixel
Arizona State University

ABOVE A topographic map of the Moon northward of 75° latitude produced by the laser altimeter on LRO. *(NASA/GSFC)*

ABOVE RIGHT A view of the Moon northward of 80° latitude produced by the main camera on LRO. *(NASA/GSFC/ASU)*

BELOW A topographic map of the Moon southward of 75° latitude produced by the laser altimeter on LRO. *(NASA/GSFC)*

BELOW RIGHT A view of the Moon southward of 80° latitude produced by the main camera on LRO. *(NASA/GSFC/ASU)*

the vicinity of the south pole that was interpreted as *direct* evidence of a water frost on the surface.

In addition, a radiometer that measured the temperature of the surface established that some of the polar craters were at 35K. This was proof that the Moon does indeed possess permanently shadowed areas, as it wouldn't be possible to maintain this temperature if the Sun shone every now and again.

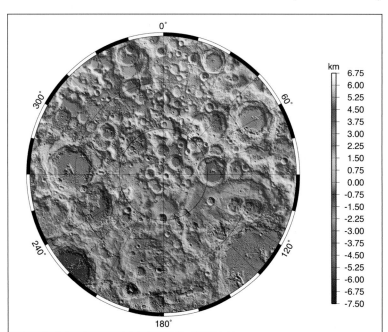

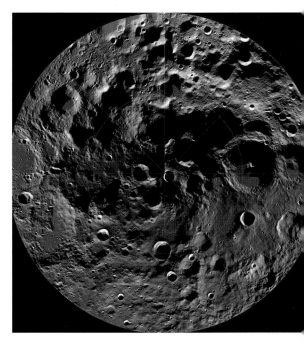

India joins in

Chandrayaan 1 achieved a polar lunar orbit on 8 November 2008. It was the first mission to the Moon to be organised and launched by the Indian Space Research Organisation.

Its suite of instruments included an infrared mapper funded by NASA. By the time the craft fell silent in August 2009 this spectrometer had scanned more than 95% of the surface.

It detected infrared absorption features that implied the presence of water molecules in the equatorial zone. The water was present in the uppermost few millimetres (by the nature of such an observation, the instrument couldn't sense any deeper than that) in the form of a thin film of molecules coating the regolith material.

Carlé Pieters of Brown University on Rhode Island was the principal investigator for this mapper. When she discussed the data with a colleague on the Cassini mission, it was realised that when the Saturn-bound spacecraft made its lunar flyby in August 1999 one of its instruments had spotted evidence of water at all latitudes. By the time the instrument was calibrated in the Saturnian system in 2004 to permit proper analysis, this lunar data had been archived for attention later.

At this point, scientists on the Deep Impact mission were asked to undertake observations when their spacecraft flew by the Moon in June 2009 on its way to a comet.

It took measurements over the period of a week in order to track the rising Sun. This time the data was conclusive: there was indeed water at all latitudes and the strength of the signature was correlated with the surface temperature. There was a strong water signature at sunrise that diminished towards noon and then resumed

ABOVE An artist's depiction of the Chandrayaan 1 spacecraft. *(ISRO)*

its initial level by sunset. The fact that the signature wasn't strengthened during the long and chilly lunar night meant the process involved the solar wind.

The evidence was consistent with Crider and Vondrak's theory that protons in the solar wind react with oxides in the regolith to create water molecules.

The molecules are lost in the mid-day heat, but their formation resumes in the afternoon. The mid-day temperature is hottest in the

CHANDRAYAAN 1 MISSION

Launched on 22 October 2008, India's first mission to the Moon entered a polar orbit on 8 November to carry out a global survey with a variety of instruments. Although it prematurely fell silent on 29 August 2009, it had by then accomplished most of its objectives.

LEFT Carlé Pieters. *(Brown University)*

ABOVE An artist's depiction of the LCROSS spacecraft following its Centaur stage on a plunging dive towards the Moon. *(NASA/Ames)*

equatorial zone and declines with increasing latitude. In the polar regions the molecules accumulate in the cold traps.

A smashing success

Meanwhile, the large rocket stage which had boosted Lunar Reconnaissance Orbiter to the Moon had used the *gravitational slingshot* of a lunar flyby to execute a large orbit of Earth that would enable it to crash into the Moon's south polar area several months later.

At that time, a secondary payload called the Lunar Crater Observation and Sensing Satellite (LCROSS) refined the trajectory in order to aim for the 100km crater Cabeus which LRO had found to be enriched with hydrogen.

In the final approach, LCROSS separated from the Centaur stage and braked in order to trail behind by several minutes.

In contrast to the grazing impact of Lunar Prospector, the almost vertical plunge of the 2,250kg rocket stage would release an energy equivalent to detonating a metric ton of TNT.

The impact was expected to make a crater 30m wide and 5m deep. It would be more than deep enough to excavate any ice in the upper regolith. Although most of the estimated 1,000 metric tons of ejecta would follow a shallow trajectory and remain in shadow, some of it would create a plume that would rise above the walls of the 3.5km deep Cabeus into sunlight.

Being just 5° from the pole on the Earth-facing side, Cabeus would permit telescopes on Earth a fair chance to observe such a plume.

On 9 October 2009 LCROSS monitored the stage's impact site for 6min before itself striking nearby. It saw the faint flash of the impact and a plume rising into sunlight. Terrestrial observers saw nothing.

A month later, NASA announced a number of lines of evidence which implied the presence of water in both the high-angle plume and the lateral ejecta.

A near-infrared spectrometer detected strong absorption bands within the plume characteristic of water vapour and water ice.

Furthermore, an ultraviolet spectrometer took a spectrum of the plume that showed an emission feature at a wavelength indicative of hydroxyl ions created by the dissociation of water molecules as they rose into sunlight.

Tony Colaprete, the principal investigator for LCROSS, described this emission feature

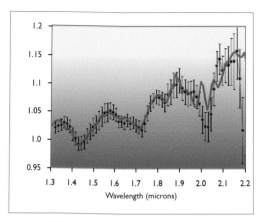

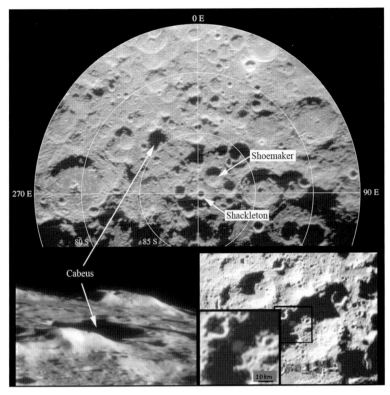

ABOVE The near-infrared spectrometer on LCROSS observed the plume from the impact of the Centaur stage. The data is well modelled by various compounds, including water. *(LCROSS/ NASA/Ames/Anthony Colaprete)*

RIGHT LCROSS dived into the shadowed crater Cabeus. Top: an image of the south polar region of the Moon by the Arecibo Observatory operating as a radar at a wavelength of 70cm. *(Arecibo Observatory and Bruce Campbell of the Center for Earth and Planetary Studies at the Smithsonian Institution)* **Left:** a near-infrared image taken using adaptive optics at the time of the impact. *(Palomar Observatory/Caltech)* **Right:** a visible-light image by the LCROSS spacecraft 20sec after the impact of the Centaur stage, showing the rising plume catching sunlight. *(NASA)*

as 'the eureka moment' that confirmed the presence of water.

If the impact excavated the predicted 30m crater, then the 100kg of water believed to have been in the plume represented a regolith fraction of 10 parts per million.

In recent years a number of missions have

LRO/LCROSS MISSION

Launched on 18 June 2009, LRO entered a polar lunar orbit on 23 June and on 15 September it initiated its principal mission employing a variety of instruments.

Launched with LRO, LCROSS remained attached to the Earth escape stage. The gravitational slingshot of a lunar flyby on 23 June placed it into an Earth orbit that was inclined at about 80° to the ecliptic and had a period of 37 days. After three orbits, it approached the south pole of the Moon on a near-vertical trajectory. Just prior to arrival on 9 October, LCROSS separated, then thrusted to delay its arrival by several minutes and observed the impact of the stage prior to itself crashing nearby.

LEFT Anthony Colaprete, LCROSS project scientist and principal investigator, shares the preliminary results from the impacts in Cabeus crater at a press conference at the Ames Research Center on 13 November. *(NASA/Ames)*

ESA'S SMART MISSION

SMART 1
Launched on 27 September 2003, it used an ion thruster to slowly spiral out to the Moon, then entered a polar lunar orbit on 15 November 2004 for a global survey with a variety of instruments.

LEFT A view of the Chang'e 3 lander taken by the Yutu rover. *(Chinese Academy of Sciences)*

LEFT A view of the recently deployed Yutu rover by the Chang'e 3 lander. *(Chinese Academy of Sciences)*

SELENE (KAGUYA) MISSION

Launched on 14 September 2007, Japan's first mission to the Moon entered a polar orbit on 3 October for a global survey using a variety of instruments, including a high-definition television camera.

BELOW A distant view of the lander. *(Chinese Academy of Sciences)*

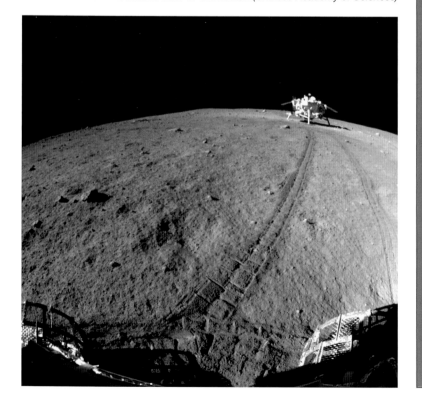

CHINA'S CHANG'E MISSIONS

Chang'e 1
Launched on 24 October 2007, China's first mission to the Moon entered a near-polar orbit on 5 November from where it mapped the lunar surface.

Chang'e 2
Launched on 1 October 2010, it entered lunar orbit on 6 October. After refining the map of its predecessor, it departed for the L2 Lagrangian point of the Earth–Sun system in order to exercise the Chinese tracking and control network.

Chang'e 3
Launched on 1 December 2013, it entered lunar orbit on 6 December and then landed on 14 December at 44.12°N, 19.51°W, where it deployed the Yutu rover.

been proposed that would land in one of the cold traps and drill for water ice. As yet though, none of these projects has been funded.

The European Space Agency, Japan, and China have all sent orbiters to survey the lunar surface, and China deployed a small rover as a precursor to a sample return mission.

Various nations, including Russia, intend to use orbiters, landers, and rovers to carry out a variety of scientific investigations.

Renewed gravity mapping

When America renewed its interest in the Moon in the 1990s, a high priority task was to better understand the gravitational field of the Moon.

Clementine was far more capable than its Lunar Orbiter predecessors. Its state-of-the-art equipment included a laser altimeter that could measure vertical heights to within 40m. It worked in conjunction with a camera which had a spatial resolution of 10m to enable the topography of the major surface features to be mapped.

It was immediately evident that there were many more basin-sized cavities than had been believed. In fact, they were so numerous they overlapped.

Although the oldest basins were difficult to see in overhead imagery because subsequent impacts had superimposed their own landforms, the altimetry revealed their profiles.

This was particularly so for the 2,500km-wide Aitken basin which lies just beyond the south pole and is only visible from Earth at a favourable libration as a chain of tall mountains. It is spectacular in the altimetry because the cavity is 12km deep, but being so battered, this enormous structure hadn't been recognised in the Lunar Orbiter imagery.

The maria dominate the visual perception of the Moon but they are distinct from the basins they occupy, and therefore are a distraction.

In order to really appreciate the magnitude of the bombardment that gave rise to the basins, it is best to view a relief map. Unsurprisingly, most

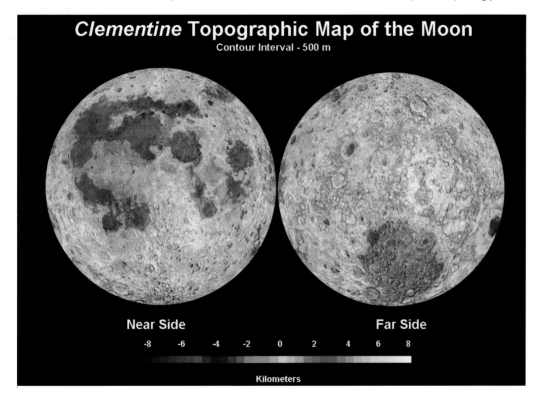

Clementine Topographic Map of the Moon
Contour Interval - 500 m

Near Side

Far Side

-8   -6   -4   -2   0   2   4   6   8

Kilometers

LEFT A topographic map of the Moon produced by the laser altimeter on Clementine. *(Clementine/Paul Spudis of the Lunar and Planetary Institute)*

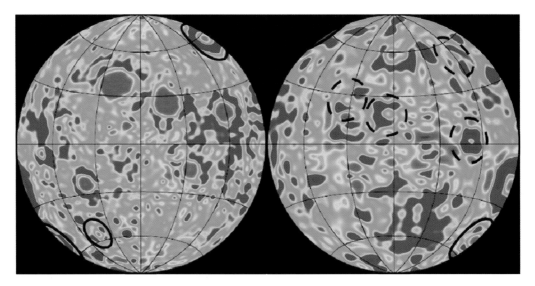

RIGHT A map of the Moon's gravity field by Lunar Prospector. *(NASA)*

of the newly identified basins were near the limb and on the far side.

The highest lunar terrain lies on the far side and the overall vertical variation from the tallest peak to the bottom of the deepest depression is 16km, which is comparable with the range from the highest terrestrial mountain peak to the deepest ocean trench. In general the lunar surface is extremely rugged, but the mare plains possess slopes of less than one part in a thousand.

By combining the laser altimetry with radio tracking, the gravity data could be processed to show crustal thickness. The results revealed the average thickness of the crust on the far side to be 68km, which is some 8km thicker than the near side.

Welcome though the Clementine data was, it just whetted the appetites of the scientists. After spending 18 months in a 100km polar orbit

Lunar Prospector descended to 10km in order to improve the resolution of the gravity map by at least a factor of 100.

The data revealed the existence of a number of previously unsuspected mascons in lava-filled craters.

Additional gravity data was gained by tracking later missions, including those of Japan and India, but better information was to come from the Gravity Recovery and Interior Laboratory (GRAIL) spacecraft operated by NASA.

Maria Zuber of the Massachusetts Institute of Technology was the principal investigator for the GRAIL mission.

A pair of small spacecraft named Ebb and Flow were launched together in September 2011. They flew paths that enabled them to enter lunar orbit 25 hours apart at the end of the year. As they orbited the Moon,

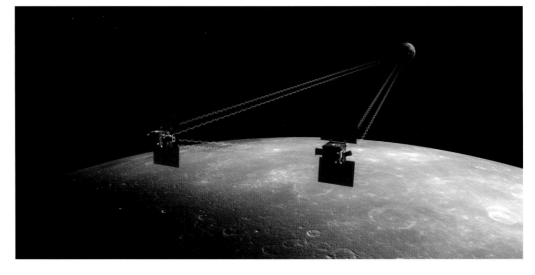

RIGHT An artist's depiction of the Ebb and Flow vehicles of the Gravity Recovery and Interior Laboratory (GRAIL) mission. *(NASA/MIT)*

the two vehicles transmitted radio signals that permitted minuscule changes in their separation to be measured and this data was used to map the gradients in the gravitational field at a very high resolution.

After conducting the primary mission from a circular polar orbit at 55km, there was a second phase at a lower altitude.

The measurements of the variations in the gravitational field were several orders of magnitude better than was previously attainable. The surface resolution of 20km was sufficient to identify fine structures in the field with features at the surface and with structures inferred to lie beneath the surface.

The gravity field preserves the record of the impact bombardment, and reveals evidence for fracturing of the interior extending deep into the crust and possibly into the mantle.

In particular, the data revealed the existence of long, linear gravity anomalies that stretched for hundreds of kilometres. These are believed to be narrow, vertical sheets of solidified magma known as dikes.

It had previously been speculated that the Ocean of Storms occupied an impact basin

GRAIL MISSION

Launched on 10 September 2011, the pair of spacecraft for the Gravity Recovery and Interior Laboratory mission reached the Moon separately, with Ebb entering polar orbit on 31 December and Flow 25 hours later. Their task was to jointly survey the gravitational field in unprecedented detail in order to study the interior structure of the Moon. Both spacecraft impacted on 17 December 2012.

BELOW A high resolution gravity map of the far side of the Moon obtained by the GRAIL mission. *(NASA/ JPL-Caltech/MIT/GSFC)*

BELOW Gradients in the gravity data from the GRAIL mission revealed the presence of a rectilinear network of dense rock structures beneath the surface which might explain the origin of the Ocean of Storms in endogenic terms, as opposed to a vast asteroid impact. *(NASA/ GSFC/JPL/Colorado School of Mines/MIT)*

RIGHT Paul Spudis of NASA's Lunar and Planetary Institute. *(Paul Spudis)*

that was created very early in the history of the Moon by the largest of asteroid impacts, but the GRAIL data revealed a giant rectangular pattern roughly 2,600km wide. This new evidence suggests this low-lying region is the result of enormous crustal rifting and settling early in the Moon's history. The low-viscosity lava issued by the dikes would have produced vast flows that buried their sources.

Crustal rifting on such a scale would appear to have occurred only in this region. The fact that the Ocean of Storms isn't a positive gravity anomaly suggests it is only a thin veneer of lava.

RIGHT A topographic map of the south pole of the Moon by the laser altimeter on the Lunar Reconnaissance Orbiter featuring the elevated terrain of Mount Malapert. The summit likely receives full or partial sunlight for 93% of the lunar year (this period takes into account the inclination of the Moon's spin axis to the plane of the ecliptic) and always has a direct line of sight for communication with Earth. *(NASA/GSFC)*

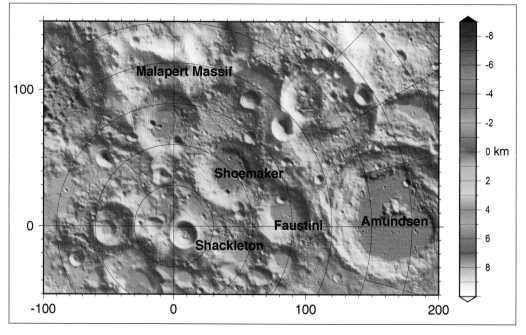

RIGHT Thermal data from the Diviner instrument of the Lunar Reconnaissance Orbiter revealed the interiors of the permanently shadowed craters in the south polar region of the Moon. *(NASA/GSFC/UCLA)*

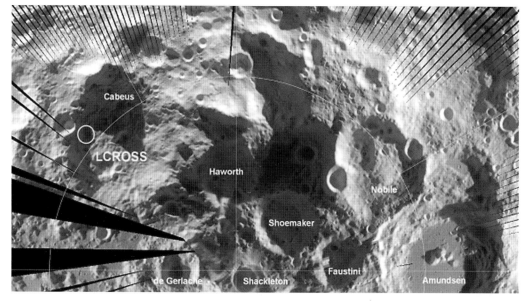

Mount Malapert

Shackleton

LEFT A view of Mount Malapert in sunlight by the high definition camera on Japan's SELENE orbiter known as Kaguya. (*JAXA/NHK*)

The future Moon

It isn't clear when humans will resume exploring the Moon in person, but when we do it would be logical to create a base at one or other of the poles.

At the press conference called to announce the Clementine results in December 1996, team leader Paul Spudis of the NASA-funded Lunar and Planetary Institute in Houston, Texas, drew attention to an elevated terrain close to the south pole that appeared to be in continuous sunlight. Calling it the Mountain of Eternal Light, he proclaimed it 'the most valuable piece of extra-terrestrial real estate'.

Not only would a base there have continuous sunlight for electrical power, the fact that the Sun circled low around the horizon meant the surface wouldn't be subjected to the thermal extremes of the equatorial zone, where the ±140°C range between lunar noon and midnight would impose severe operational issues.

The regolith is predominantly oxides of iron, titanium, and aluminium. Heating regolith in a furnace could liberate oxygen and yield raw construction materials. But the key would be water. Whilst it would be hard to 'harvest' ice crystals that were sparsely distributed in the regolith in a cold trap, any water that was extracted would reduce the amount that would otherwise have to be transported from Earth. In addition to serving in a vital life-support role, water molecules could be broken into their constituent hydrogen and oxygen atoms for use as rocket propellants.

If the helium-3 isotope delivered by the solar wind has accumulated in the regolith, then Apollo astronaut Jack Schmitt has advocated mining this for use in fusion power reactors. But that isn't likely to happen until many years after a lunar base has been established.

The prospects for mankind as a spacefaring species are encouraging. The solar system is our backyard. It has all the natural resources that we'll ever need. In energy terms, low Earth orbit is halfway to anywhere. The resources of the Moon will enable us to venture into deep space. As with the Apollo lunar program, it is simply a matter of deciding to go.

BELOW The energy requirements for travelling between various destinations in the near solar system. (*Woods*)

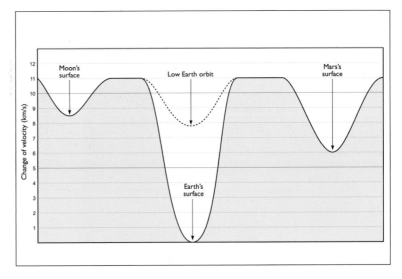

Postscript

O n 31 December 1999 National Public Radio in the United States asked Sir Arthur C. Clarke, renowned for forecasting many of the developments of the 20th century, whether anything had happened in the preceding 100 years that he never could have anticipated. 'Yes, absolutely,' he replied, without a moment's hesitation. 'The one thing I never would have expected is that after centuries of wonder and imagination and aspiration, we would have gone to the Moon ... and then stopped.'

Yet, in retrospect, our action made a certain sense. The Apollo missions were in one sense an element of the 21st century that was, rather amazingly, undertaken using 1960s technology. It would have been very difficult to expand that effort into a permanent presence. Certainly by the turn of the century we could never have attained the kind of facilities portrayed in Stanley Kubrick's 1968 movie *2001: A Space Odyssey*, which was both inspired by one of Clarke's short stories and was made under his guidance.

When the Apollo crews walked on the Moon, the longest that anyone had spent in space on a single flight was a fortnight. To advance further, we required time to develop better technologies and acquaint ourselves with living in the space environment.

The International Space Station allows us to establish the capability to live and work in low Earth orbit. In energy terms, low orbit is 'halfway to anywhere', and the solar system, as the 'backyard' of the human race, contains all the natural resources that we will ever need. All we have to do is to develop the means to exploit them and become a genuine spacefaring species.

The Moon can be reached in a few days of travel and its surface is rich in minerals which are essential to a high technology civilisation. By building a base of operations on the Moon we shall be able to learn how to 'live off the land'. And the Moon, with just 1.23% of the mass of Earth and 21% of its escape velocity, has a much shallower *gravitational well* that makes it an invaluable 'stepping stone' to destinations beyond.

So there we have it!

The Moon is humanity's future.

LEFT A view of the Moon from the International Space Station. *(NASA)*